The Magic of the Farm

For my dear parents, Bernard and Elaine

and

my Aunt Eleanor

and

my second cousin Dave Rowell

The Magic of the Farm

Seven Farm Stories

*A*dvantage
BOOKS

PHILIP ALEXANDER URIE

Illustration by RODERICK ALEXANDER WELLS

Published by: ADVANTAGE BOOKS™, Longwood, FL
                  www.advbookstore.com

Library of Congress Catalog Number: 2023947889

Name:	Urie, Philip Alexander, Author
	Wells, Roderick Alexander, Illustrator
Title:	*The Magic of the Farm*
	Philip Alexander Urie
	Advantage Books, 2023
Identifiers:	ISBN Paperback: 978159757641, eBook: 978159757740
Subjects:	RELIGION: Christian Life – Inspirational

Revised Edition: April 2024
24 25 26 27 28   10 9 8 7 6 5 4 3 2

The Magic of the Farm combines the rural flavor of James Rebanks' *Pastoral Song: A Farmer's Journey*, (Custom House 2021, UK: *English Pastoral*) and the farm world of James Herriot, with the dry humor of northern Vermonters. *The Magic of the Farm* gives presence and face to this region like Jim Sterba's *Frankie's Place* (Grove Press 2003) does for Maine's Acadia. Seven related stories include two extended conversations and cover three generations of people who grew up on small dairy farms.

Philip Alexander Urie

Table of Contents

1

The Magic of the Farm

The Magic of the Farm is a fictional family conversation set in 1933 based on themes found in the many pages of written recollections of my mother, Elaine (Alexander) Urie, and her sister Eleanor about the Aldrich family farm in Glover, Vermont, where they grew up, for a time at least. It was the first farm south of Glover village. They wanted to produce a book, as evidenced by several chapter titles and the fact that their writings were stored together. A few of their recollections follow:

"The depression came early and stayed late in Vermont and affected almost everyone in the small towns. . . In Robert Frost's poetry there is and was the magic of the farm. . . Grampa was pleasant to be around, playful and we felt his love for us . . . a small man in stature, but we felt he was grand. . . Daddy played his violin for us and visited people and fixed engines all over town. He was a genius, I heard them say. . . Mother took care of Grampa Alexander during his last years. He was very sick and with the six of us I wonder how she managed. . . The garden was a real project and mother planted and did most of the work. She probably worked harder than everyone else all put together. . . Her teacher said she was a 'natural', that her touch on the organ was perfect. . . Richard was our pet. . . The large teakettle was always singing and reminded someone to fix the fire. . . How we loved it, being together . . . Our house always warm by mother's touch and a dad who cared for us all; this friendly place we called 'home'. Warren 1923, Eleanor 1924, Wayne, 1926, Elaine 1927, Rebecca 1929, Richard 1933."

The Magic of the Farm

"Now you all look out for each other, and be back in half an hour. Remember we're picking berries and beans all day today, my little dears."

So off they went down the driveway and across the bridge with Wayne holding Elaine's hand and Warren carrying little Rebecca.

Looking up at Warren, Becky asked, "Why do they call our road "Lover's Lane"?

"Because this is where people go to kiss."

"Why don't they go home and kiss?"

Elaine giggled and Wayne and Warren both smiled at her.

"It's more exciting over here."

Becky thought a minute as they walked along and said, "It's like magic here!"

Walking carefully between the smooth dark gray boulders and rocks, they all went down to the river. In the summer the river was quiet enough so they didn't have to yell to be heard.

"Kiss me", Rebecca said to Wayne.

He gave her a kiss on both cheeks.

As they splashed around Warren thought to himself, "I don't like winter. Someday I'm leaving Vermont and moving south."

"Is there room for one more?" Eleanor asked from up on the bridge.

"Did you finish your book?"

"I have only two chapters left." She thought how she would like to write a book someday.

"You read a lot, like Daddy."

"It's hot in the attic. Then I remembered I've only gone swimming once this summer." So down the bank she went. Eleanor hugged her sisters, pulling them close. We better not stay too long. Mamma's already picking beans."

The boys helped everyone float small sticks called "boats" in the river.

Elaine said, "Mamma says I eat all the raspberries."

"Well, just be sure to pick more than you eat, and mamma won't scold you."

"You know what I love second best?"

"What?"

"Raspberry pie. Mamma always makes three pies for Thanksgiving, but someday I'm going to make eleven different pies and eat one piece of each kind with a cup of Mamma's favorite tea. And I'm going to have all my friends over."

Before long it was time to go. Wayne sang "*Let Me Call You Sweetheart*" and "*The Old Gray Mare*" on the way home. He was so good at remembering the words to songs.

Warren, Wayne, and Eleanor went over to Mamma and asked where to start.

Elaine and Becky ran to Grampa sitting on the porch where baby Richard was asleep beside him. Elaine glanced at Richard and felt such love for him. Then she remembered how Wayne put the old basketball, two baseballs, and his baseball glove beside him in the crib when he was not even a day old. Everyone knew Wayne was thinking of a baseball team!

"Buppa, you said Jesus loved a family just like us. Tell us their names again."

"It was Mary and Martha and Lazarus – two sisters and their brother. They were all about the same age – like you. Jesus loved them all. Now, shouldn't you be picking beans?"

"Bye Buppa. We love you, Buppa."

Mamma gave them each a woven basket and they began picking near the others.

"Is Grampa going to die?" Elaine asked her brothers.

"Everybody dies" said Warren.

"His hands shake and sometimes he stutters when he prays."

"He's always praying on the inside" said Wayne.

Eleanor added, "Don't think about it. I don't think he'll die for a long time."

Wayne sang song after song until the beans were almost all picked.

"Joseph Warren, help me carry all these beans into the kitchen. Wayne, you and Elaine finish picking the last row of beans. And Rebecca, you help Eleanor pick all the raspberries. Then come inside when you are done."

The water was almost boiling. Mamma moved the two big pots to a cooler spot on the stove.

"I hope we can put up twenty jars today and maybe that many again tomorrow morning."

Warren noticed Mamma's hair was extra curly today and her face was sweating. But she was still smiling, and that made him relax.

"Start cutting."

Warren thought of all the mothers and their children doing exactly what they were doing. It was important to hurry and get the job done.

"Are you still liking school?'

"I like history and spelling best."

"I hear you like to tell yarns."

"What's a yarn?"

"A long story that has lots of funny parts."

"I don't tell stories, Mamma. I like to talk about real things that happened. I make people laugh, and it makes me happy."

"It shows in your face. That's a good thing, Warren. You probably get it from my grandfather Charles Dickens Hyde. He was a professional newspaper writer, sort of like Mark Twain."

"But people laugh at what I say, not what I write."

"Then maybe you get it from listening to the Victrola."

In only ten minutes they had enough beans cut to make the first four jars. Mamma placed handfuls of beans in the steeping hot water and swirled them around several times with a big wooden spoon. After two minutes she used tongs to fill the jars with the beans along with hot water and salt. Then they began the process all over again.

"Tomorrow when we've finished, we'll take all these jars to the cellar. The yellow beans are going to taste so good this winter. Now what do you say you and I have a doughnut and some cocoa."

Soon Eleanor came walking toward the house with two pails of raspberries. Becky had red lips. But before going into the house she wiped her lips on her sleeve. Grampa smiled at her and touched her lips.

"How many raspberries did you eat, little girl?"

Becky didn't dare to say a word.

"You both sit up here and have a bowl of raspberries sprinkled with sugar. You've earned it. Tomorrow we'll make jelly out of the rest", Mary said, as she hugged Eleanor.

Just then Joseph came in from outside and announced, "Nola had her calf this afternoon. She never made a peep. Her udder is as big as twenty watermelons!" They all laughed.

Eleanor begged, "Can we keep it, Daddy?"

"No, it's a bull."

"What's an 'udder' ", asked Becky.

"Look it up in the dictionary."

"You know she can't read yet. Becky, the udder is where the milk comes from."

"Oh, you mean 'bag'."

"Yes, most people call it a 'bag'."

"Warren, we're butchering 'Pluto' tomorrow morning right after breakfast.

"OK, we'll both help."

Becky begged, "Can I help?"

"As long as you don't get in the way and you wear a warm coat."

Wayne came carrying two more baskets of beans heaping full and set them on the porch.

Daddy, can we go see Nola now?"

"No, let them rest for a while. You can see them when we bring the cows in. I promise. We've had a big day and we should all slow down."

"Eleanor, help me peel potatoes, and I'll do all the rest."

"Daddy, first tell us again how you and Mamma played music together before you were married."

"Yes, well, do you remember who Roy was?"

"Mamma's good friend?"

Philip Alexander Urie

"Yes, from up Sheffield way. And he was my friend too. He let me play his violin until I got my own."

"Was he any good?"

"Mamma thought so." Joseph winked at Mary. "He taught me a lot about playing the violin and he let us borrow his sheet music. Sunday after dinner I would walk up the cow path, go through Mr. Perron's woods and up Clark Hill all the way to the house where your Mamma grew up. She played the piano while I played the violin. Sometimes Roy showed up too."

Becky asked, "Did you kiss her?"

"Did you have to knock on the door or did Mamma let you right in?" asked Wayne.

"Most of the time your mother was looking for me, but a few times I had to knock. Then suddenly it seemed like home just to be with her. So yes, I kissed her right in front of everybody. When it was time to go she whispered something to me just outside the door."

"What did Mamma say, Daddy? Please tell us."

She said, "You have 'kissy lips' ".

From the porch Grampa could hear everyone laughing.

"Someone else said that before I did!" said Mary.

Mary and Joseph looked romantically at each other as if they could float away somewhere together.

"Now let's all take a nap or read or do something quiet."

Warren went upstairs where the sunshine was brighter. He found the Aldrich family's old leatherette book "The History of England" to read and lay on a blanket in the attic. Eleanor curled up in the corner of the couch dreaming of the day when Aunt Edith would come for a visit and teach her how to dance the Charleston. Wayne sat on the couch reading the installation instructions for Dr. Buck's new oil pressure gauge. Elaine and Becky fell asleep on each side of baby Richard. It felt good to be all bundled up in blankets and to watch Grampa smile. Except for the kitchen it was quiet everywhere, at least for a while.

"Listen! It's Uncle Al!"

Each child ran out just a few steps apart and down the steps on the porch. The siren was louder than the motorcycle. Eleanor hugged her uncle around the neck and blew the big silver whistle that hung on his chest. He hugged and kissed them all and made sputtering noises, mimicking his motorcycle as they all took turns blowing the whistle.

"I've been to Danville and Greensboro Bend and Sheffield Heights, but I haven't seen any kids as cute as you!"

He kissed and hugged them all and gave each one a piece of peppermint candy.

"You're looking good in that uniform, Alec", said his brother Joseph.

"I'm now an official 'Uniformed State Trooper'. Can you believe it? And we'll soon be carrying big pistols. I hope I never drop it on my toe!"

"Come in for coffee and a doughnut."

"I just can't. Flossie and I have big plans! But I just had to see my little darlin's."

"I understand."

"How's Daddy?"

"No worse today."

Erwin went over to Edwin and gave him a long hug and a kiss on the cheek.

"Sorry I can't stay longer, Daddy."

"It's alright. True love does something to men, you know! Come back soon. The children just adore you."

"See you soon, Daddy."

Edwin got up and shuffled to the barn. Elaine followed not far behind.

"It's time to bring the cows in."

Edwin rested his elbows on one of the window sills on the end of the barn and called out over and over again, "Come boss, come boss".

His words echoed off the hills and to Elaine it seemed like it would last forever.

"Buppa, do you miss Gramma?" asked Elaine.

"I loved my Cena, I mean Francena. She was a hard worker. But we surely had a lot of fun. Cena was like magic to me. She thought I was a real prankster! You remember how tall she was compared to me, don't you?"

"Yes, and she always said to me, 'I don't ever want you to use swear words.' "

"That sounds like my Cena. Always be a good girl for Jesus, Elaine.

"Sometimes I'm afraid when I hear you bang on the pulpit with that gavel, Buppa."

"Don't be afraid, Elaine. The Bible says 'God is love'. I'm just warning people about the Judgment Day. You know Jesus is coming back, don't you?"

"Yes, Buppa."

"He'll gather us like sheep, and we will live forever."

By this time the other children had arrived at the barn to see Nola's new calf. Edwin turned a pail upside down and sat watching each of his grandchildren work. Soon they had all the cows hitched up. Wayne and Eleanor worked together to get the silage down from the silo while Warren shoveled sawdust under the cows. After petting the calf for a long time they heard Mamma ringing the dinner bell. Everyone took their turns washing their faces and hands, combing their hair, and settling into their usual places at the table. As usual Grampa prayed.

"Are you and Mama going to play together tonight?"

"Right after chores." Joseph stared at Mary and hinted, "Frederic Chopin's Nocturne?"

"We all like that one so very much. I could hear it ten times!"

When the milking was done and the cows were put back outside for the night, Joseph left the generator on so there would be lights on in the house for their little concert.

"Daddy and Mamma, you look so good together."

Soon it was time for bed. After a lot of hugging and kissing, off they went in a bunch to their rooms upstairs with their one small lantern. Later that night a few words were spoken.

"I'm cold."

"Get in", said Warren as he lifted the blanket up high.

Elaine knew her brother loved her. It was magic. Fifty years later she recorded these words:

"When my grandparents Edwin and Francena (Aldrich) Alexander could no longer do the work on the Aldrich farm, my parents went there to live. Joseph and Mary (Stone) already had three children. Wayne was

mother's prettiest baby. She told me. His blonde hair "almost white" and blue eyes were winners! (Mom always said she was "the black sheep of the family" because she was the only one to inherit her mother's dark eyes.) *I was the first baby to be born on the farm – named "Elaine" from the story of "King Arthur's Court" as my grandfather Edwin's middle name was "Lancelot".*

My favorite pastime can take me back to the old farmhouse – where we lived all together for awhile at least. It was a real treat to hear all about the people who lived in this house before us.

Mother played the piano for people to sing – then six little children gathered around the piano with Dad on his violin. We sang the old songs. Sister Eleanor was so nice to us all and baby Richard so dear. Big brother Warren helped, doing real chores. My buddy Wayne soon became a real helper too. Becky grew to become my best friend. She and I slept in the same bed.

Mother was very busy from sunup until she dropped to sleep. She told me "the broom and dustpan were in the corner". Her days were work-filled and the kitchen alone a world in its own. The big black iron stove used for generations before was almost always going with the wood to fire it. The top was always cooking potatoes, meat and vegetables and the flapjacks on the very old griddle which turned one at a time. Canning took hours on end.

I remember so well, the big kitchen table was in the center of the room and used for our meals and all other activities. This is where we cut and pasted pictures, colored, and played games. Mother hung a sheet in a doorway so we could play shadow games. We sang songs while she played the piano and Dad on the violin. We were never spared anything it seemed.

The big iron sink was in a corner where two metal dishpans were used. The old teakettle made hot water for the dishes. Mother had a piece of cloth and this was the dish rag, usually a square of old long john material. There were no purchased towels and sheets were made along with diapers for the babies.

There was a large bedroom up over the kitchen, and one day our mother said, "This is just the kind of a day Ma might come." As she *spoke they did appear at the foot of the hill. They probably came by train and how they arrived at the farm I don't know. Blanche and Henry "worked out" for people - the last time in New Jersey. We all piled onto Grandma Stone's bed to hear about New York City!*

They sent us very large containers of honey. It wasn't clover honey! Grandma Stone did a lot for us as she was very clever. She made quilts and some of our clothing.

There was a pasture hill to explore where we picked flowers and strawberries and tipped over stones to watch the ants. We roamed the hill along the cow paths, naming them Glover Street and Sheffield Heights, Runaway Pond Road, and others. It was a game for us.

The barn was another whole world to us, with the cows, calves, and pigs underneath the stable. The horse barn was on the ground level with work horses including Kate the white horse. She took us to town by wagon or sleigh and knew the farm turn, even though she was blind. We also had a collie dog who came to the school yard and waited for us to come out! From the main road Kate came into the lane and across the bridge over the river and on up the hill to the farm. Coming home from school the lane was so cool. On Christmas mother put sleigh bells on the harness when we went to the village for the church program.

I will never forget the spring that Grampa (Edwin) died. She set us at the table and said to "play cards". Grandpa was embalmed in the parlor room. Dad helped and after awhile we were told to go to the room and see him in the casket. Then we went up to bed.

The house was lit by candles and oil lamps. The rooms were dark at night. We usually found our way around in the darkness! The house was cold in winter. We took jars of hot water and heated bricks to our beds. I remember how good it felt for awhile.

How we loved the years at the farm being together – this friendly place called "home" – our house always warm by mother's touch and a dad who cared for us all."

The Magic of the Farm

**Joseph and Mary Stone Alexander in Maine 1950
Courtesy of Joan Alexander**

2

Gramma Joe's Smile

When friends and neighbors from here in Pennsylvania ask me where the best places in Vermont to visit are, I ask them a question: "Do you want to see Vermont or would you also like to meet some true Vermonters?"

I explain that, for me, real Vermonters can only be found in the northern part of the state. ***Gramma Joe's Smile*** tells something of the life of my Gramma Alexander, a true Northeast Kingdomer. Mary Elizabeth Stone was born in Albany, Vermont and lived most all of her adult life in Glover. To me these two towns along with Craftsbury and Greensboro make up the real "Northeast Kingdom". These four towns with their many lakes make up beautiful country with consistently striking views occupying each mile of road. Their small villages have no side streets, only the "main drag". Calling the East Albany area "the kingdom" was popularized by Gramma Alexander's father, Henry Stone. "Kingdom Road" is still there today.

Henry would span the horizon and say with the greatest of admiration and happiness to be here: "Oh, the kingdom!"

Though it is usually 10-20 degrees colder than the rest of Vermont due to its high elevation, the Northeast Kingdom (NEK) is a very unique part of Vermont, especially around Andersonville in the town of Glover and its two nearest neighborhoods – Mud Island to the east and South Albany to the west. The villages and farms have their own and more distant mountains and ridges as their backdrop, some over 50 miles away. Just as Vermont fancy maple syrup is best described as "romantic", the high elevation views of the kingdom are truly "delicious". One visit to this high country and one longs to see it over and over again!

Gramma Joe's Smile

We were the only cousins who called our grandmother "Gramma Joe"! She wanted it that way. When our oldest sister Kathy was only three years old, someone asked her about her grandfather.

"How's your Grampa Joe?"

Kathy thought this person had asked, "How's your *Gramma* Joe?"

For her whole life "Gramma Joe" was all the five of us ever called our Grandmother Alexander.

I was perhaps ten years old when my cousin Edwin Alexander asked me, "Who are you talking about when you say "Gramma Joe"?

I thought he had lost his mind! But of course he only knew her as his "Grammie Alexander".

Grampa Joe worked at Urie's Garage in Barton. When he died in 1951, a year before I was born, Gramma Joe found a smile to wear. Somehow it became part of her. It helped to have Elaine and Bernard and their five children living so near to her.

We loved to go down to the Barton Restaurant to see Gramma. We went there to pick up our sister Kathy at 7 pm. Gramma Joe's smile was always there to greet us at the side door. All her customers had to step down off the sidewalk to enter the front door, but we got to go through the kitchen door on the side. The gas stove, big and black, was always boiling or frying something, just like her farmhouse stove in Glover.

"Mumma, your burgers are too fat", commented Mom.

Gramma ignored her words, because her customers kept coming back. She quickly peeled a potato and put it in the "cruncher". Elizabeth and I smiled as the fries boiled in the fat. We left Mom's side to search out song titles on the flapping pages near the juke box in the back. She and Gramma needed to talk anyway.

"Does that soo-chee?" Gramma asked, as she delivered our French fries.

They suited us just fine. Gramma knew we wouldn't answer because we were too busy eating. Gramma smiled and went back to her duties in the kitchen, while Liz and I ate and watched Kathy be a waitress.

I wish we had a photo of Gramma Joe working at the restaurant. But she never liked to have her picture taken. She didn't think she was photogenic.

She always said, "Put that camera away! I don't take a good picture."

But there are several good pictures of her. In the good living room Mom had a large oval picture of her mother, little Mary Stone. Then there is the picture of Gramma and her sister Bessie seated together on an open car in their "courting days" on Clark Hill.

During the "Depression Years" Gramma Joe was always busy providing for her six children. Mom says she remembers her putting up 500 quarts of vegetables each summer. And like most families Gramma had potatoes and meat going on the stove all the time. For a break she and the youngest children went across the brook and up through a little path in the woods to visit with the Gaboriault farm family across from Westlook Cemetery. They only spoke French at first, and so they all had some trouble communicating with each other! But it was a way for Gramma to sympathize with, and to laugh and smile with another young mother who also knew what hard work was, in the house and barn and on the whole farm.

It was not until 1823 that northern Vermonters and southern Quebecers knew what country they were living in. The two governments simply did not settle on where the border would be. So once decided upon and declared many Quebecois found themselves and their citizenship in Vermont, just as many Vermonters found themselves permanently living in Quebec!

After Grampa Joe died, I recall traveling to Quebec to visit his father's relatives in Mascouche. Those visits with the Gaboriaults helped Gramma to talk with the French-speaking relatives. At least Gramma Joe seemed better able to talk with them than my parents!

Gramma Joe's second-floor apartment was on Glover Street in Barton, two doors south of where the Crystal Lake Falls runs under the street and into the Barton River. We called it the Hagar house. Twice each week Mom, Liz and I would go visit her. We loved to pull the hall light on with the string and run upstairs to her door and knock.

She opened the door and said, "Hello, my little darlin's."

Gramma had to carry kerosene up the stairs every other day. She had no running hot water or central heat. Yet it was a real home with photographs to look at, games to play, and Canada Mints to eat.

"Isn't that something!" Gramma would say when Mom reported how well we were all doing in school.

We were all very good "spellers". I believe we inherited it from our Gramma Joe. But she did the cryptogram in the *Newport Daily Express* every day. This amazed us! She gave us clues, but we simply could not do them.

One time Gramma was eating something she called "Salmon Pea Wiggle". We thought she was joking! It was hot milk gravy with chunks of canned salmon and peas poured over crushed crackers. We learned to love canned salmon because that's what our Gramma Joe liked. Later we learned how healthy salmon is and how to fry salmon patties.

Sometimes when we got ready to leave, Gramma would try to force two dollars on Mom. Mom would refuse to take it. But Gramma refused to keep it. After a little battle with the money going back and forth between them, Gramma Joe's smile finally won out. She would give the two folded bills to Liz or me, and off we'd go with a good hug from her.

Before arthritis stiffened her fingers, Gramma Joe loved to play the piano in our farmhouse in West Glover. She played the organ at church too. Aunt Harriet described her as a "grand lady". At our house she played very loud, rollicking tunes, singing a few lines here and there as she played. Kathy, Liz, and I tried to sing a little too. But we couldn't hear ourselves or keep up with her, since Gramma played quite fast and very loud, and without stopping. It was a wild happy time for the four of us! But she did take time to teach us how to plink out a few tunes.

When Gramma would come to our farmhouse in the early spring she would go out into the backyard and look for dandelions. Dandelion greens are a Depression-era food that tastes almost like spinach. Gramma would grab the biggest butcher knife she could find and dig up the plant, including the leaves and some of the root. She scooped them all up and carried them in her apron onto the back porch. Then she went inside and got a pail of water to wash them in. Liz and I noticed there might still be some dirt and stains left on the plants even after Gramma had washed them.

We quizzed her: "What about the dirt on them?"

"Oh, don't worry about that. We'll steam them. Everyone eats a peck of dirt before they die anyway!" she said with a smile.

Once Dad and Mom decided to go on a little two-day vacation by themselves. Gramma Joe agreed to stay with the five of us. Since we never had locks on our doors, Mom told Gramma to stick a table knife against the door into the door casing each night. Then she would be safe inside with her five grandchildren.

In the late evening of the second day of Dad and Mom's little vacation, there came a soft knock on the back door. None of us heard it. More knocks followed. Then there came louder knocks on the front door. Dad and Mom couldn't stir a soul! Dad began turning the door handle and wiggling the back door. A knife fell to the floor and they were inside the dark house. They wondered how they got in through the locked door so quickly.

"Come over here, Bernard" Mom said quietly.

They both chuckled as they looked at the front door. Gramma had put the knife between the door and the casing sticking straight out instead of straight across!

We all joked about it in the morning. Our Gramma Joe just smiled.

One time my sister Elizabeth brought Gramma Joe up to our home in Swanton, and let her knock on the door all by herself.

Fran answered the door and said to a smiling Gramma Joe: "You've caught us at a bad time! The house is a disaster!"

But she wanted to see her great-grandchildren in their own house in the midst of farm country.

When Franny described to my mother how Gramma had surprised them, Mom explained that Gramma simply wanted to see her great-grandchildren as they really are.

Mom said, "Mumma likes that lived-in look."

I'm sure our home reminded her of her mothering years and her six darlings on the Aldrich farm in Glover. Elaine described her mother at our Gramma Joe's funeral:

"Back then mother had a large vegetable garden. She set out the tomatoes and planted the garden – hoed and weeded it and put it all in jars for processing. We all helped and learned so much that we all like to garden. It was not all fun and games as these vegetables were absolutely necessary for winter meals. Mother made wonderful boiled dinners, and I can still see the meat on the platter with potatoes, carrots, beets and cabbage. Our mother dressed off chickens, canned the meat as well as the beef and pork. She made the butter, the bread, the biscuits, and the maple spread, and the lovely pink strawberry frosting for the fresh cake. There were jams and jellies, the pies, apple dumplings and strawberry shortcake; and good hot meals of potato, meat, and gravy.

I saw her make the candles and the soap. She fed the hens, the geese, and the pigs.

She tended the children and the neighbor's children. She made cottage cheese and homemade ice cream and helped at sugaring time.

There were fires to be kept, endless dishes to be washed, and clothes to be scrubbed and hung out. She cleaned and filled the lamps, cared for an invalid parent, and mended and patched clothing.

She herded us in and out of school, played the piano for us to sing with, took us on picnics, and played games with us. She hitched up the horse and took us to church and Sunday School, heard our prayers, guided us and protected us, and loved us as we loved her - our mother."

Gramma Joe was always interested in people's news and spoke well of everyone, especially her 30 grandchildren. I experienced such a feeling of loss and loneliness when she went to live with Aunt Becky in Colchester in 1965. She lived there for over 20 years. Whenever I went to visit Gramma Joe, she always insisted on serving up a cup of coffee and a doughnut with a smile. It reminded me of Gramma's "restaurant days" back home in the Northeast Kingdom. But Mumma remembered serving doughnuts to her little ones in the Aldrich farmhouse with the same smile.

The Magic of the Farm

Elaine and Bernard cut their cake December 29, 1945

3

It's a Small Hotel

Mom ran her farmhouse like *a small hotel*. It surprised us to see that Mom had a pair of baby sandals clipped to the wooden clothes rack on Mom's back porch. Fran took them down and hung the baby shoes by her bed at the nursing home. Elaine loved babies!

I can see her now . . . washing the milk pails wearing a kerchief on her head . . . marveling at the tall hollyhocks by the barn and the tulips by the road out front . . . talking with Eleanor in the driveway and cherishing her long phone calls with her sisters Eleanor and Becky . . . making Dad so happy when she climbed Lone Tree Hill at the age of 75 . . . and telling us she is praying for us. How thankful we must all be, knowing our faithful Savior Jesus Christ prays for us exactly right.

It's a Small Hotel

Elaine looked for a piece of wood that would do the job.

"Bernard, I need a piece of smooth wood about a foot-and-a-half long."

"How wide?"

"Ten inches, I suppose. I'm going to paint the words, 'It's a Small Hotel' on it."

Dad cut Mom just what she needed. So she painted it and added a few flowers. By the time she was done Dad had found a bracket to hang it on. And so the sign was hung on the front of the house for all to see.

Several regular guests and relatives stayed overnight in their farmhouse, though not often in winter. They were usually there for dinner, or rather "supper", and for breakfast, often being served sugared raspberries from her red glass bowl. Over and over Elaine's children came and went, filling most every room, like a good hotel. Other people filled the comfortable rooms too. Her children brought friends and fiancées for long visits. Nieces and nephews felt welcomed to come and sit and talk.

Mom could often be heard to say: "Take your pick of beds. There are four bedrooms and five beds upstairs. You're welcome to the downstairs couch too."

Eventually Mom and Dad had their own bedroom and bathroom downstairs.

Four red-headed girls became such close friends for Dad and Mom. They were like family and stayed overnight many times. Everyone enjoyed the sound of the splashing brook down over the bank and the taste and smell of a warm farmhouse breakfast. A neighbor boy, his sister, and their mother visited often. How he loved Gramma's plain bread and butter. Sometimes he ate four slices!

Mom's cousin Wes often sat by the front window on summer mornings, sometimes detailing to Dad and Mom what it was like working for President Nixon "behind the lines" during the Cold War. Wes's relatives and their friends from Ohio made regular visits too.

Mom had a full-size electric organ and could play quite well even though she never had a lesson. In 1978 she and Dad travelled all the way to Springfield, Vermont to hear Boots Randolph play his saxophone for his "Yakety Sax" concert.

For several years Mom and Dad sponsored a neighborhood pie party. Mom always made seven pies ahead of time and froze them. The day of the party the back lawn was full of people and their pies. So there were always at least twenty pies to choose from. The farmhouse seemed more like a big hotel now. Every guest had at least one piece of leftover pie to take home. Uncle Edwin always came back the next day for several pieces of leftover pie. This is what a welcoming small hotel is all about!

Jesus spoke like this: "In my Father's house are many mansions; if it were not so, I would have told you. I go to prepare a place for you. And if I go and prepare a place for you, I will come again and receive you to myself; that where I am, there you may be also." (John 14:2-3)

Mom wanted everyone to feel welcome with her little hotel sign. But God is the best hotelier! Our graves are only temporary beds for our bodies. Far from suffering and the blackness of hell, our travelling souls have heavenly rooms to stay in when we die, just the way Jesus described them, and all at no charge! Free! Just like Mom's small hotel!

Northern Vermonters resign themselves to certain things when it comes to "expenses". "Careful" is the best concept when it comes to money. For instance: keep your money in an out-of-town bank. Bankers gossip about everybody. So be smart and let them talk in the next county over, not in your own.

Uncle Walter made quite an impression on me when he said "I don't know why people don't pay their bills when they come in the mail." In other words, why leave a bill lying around?

Fran's mom said to always throw meat to the dog with the biggest mouthful of teeth!

The main idea is to stop your money from ever dwindling to nothing.

Mom said often to us: "Keep your checkbook fluffed up."

But the question was, "How do I keep it fluffed up?"

We learned by carefully watching the way our parents lived. Every time they went shopping at the store or at a yard sale they competed to see who would spend the least! They never bought a brand-new car or built a new house. They never traveled too far and never went on a real vacation. Northern Vermonters don't take long showers and always turn the thermostat all the way down at night. They always wear socks and a hat to bed. The first time we slept in West Glover as a married couple at Christmas time Mom handed out new knit hats to Fran and to me.

"Put these hats on your heads and leave them on. We turn the heat down to nothing at 8 o'clock!"

The fight against the cold takes place both outside and inside the house. But Northeast Kingdomers never leave the electric blanket on all night. To save money they leave it on just long enough to warm the mattress a bit before heading for bed. Northern Vermonters like Dad and Mom know the mattress does not warm the man; it is the man who must keep the mattress warm, or at least the portion of the mattress he lies on! Electric blankets are a real help though.

Uncle Edwin started one of his letters with these few words: "Cold in your beds?"

Men in the Northeast Kingdom wear their pajamas until they are paper-thin. Their wives remove and "turn the collar" on their favorite shirts. When food prices go up, they simply eat less! These are some of the ways northern Vermonters keep their checkbooks fluffed up. But they happily spend money when it is absolutely necessary. So two things they never run out of are maple syrup and firewood. They also never run out of "free" hot water. The teakettle is always brewing on the wood stove in the kitchen, ready to provide the hot water for doing dishes. Of course, a hot teakettle also adds moisture to the dry air of winter.

When they realized the Andersonville Cemetery in Glover was filling up fast, Dad and Mom purchased the last six grave plots. Fran and I just looked at each other when we heard what they had done.

Then my sister Elizabeth asked, "Who are they all for?"

Mom was always good with a quick comeback, and replied without missing a beat, "Whoever gets there first!"

Northern Vermonters believe in "mud season" and would never dispute that northern Vermont has five seasons. They have nothing against dirt roads except in mud season and know there is no point in washing their car until summer. They never go swimming unless it is 98 in the shade. So they never swim for long, if at all. Northern Vermonters like Mom don't want to miss anything, and so they rarely leave the state. They don't like to stay overnight anywhere. They know the month of May could be no better than April and that their whole garden may fail because of too many rainy days in a row, all because of the cold Atlantic. They know to leave their car overnight facing the east on the worst winter days. This way if the car won't start, they can lift the hood and let the sun warm up the engine, though they rarely bother to back their car into the driveway in warm weather. They always think twice or three times before blowing their horn. Northern Vermonters rarely hit a deer or a moose because they drive slowly everywhere they go; rather than going out at night they will wait for morning's first light. They don't like the thought of getting "clipped" on the road!

Most Northern Vermont mothers devotedly cut the men's hair, and often their own.

I can still hear Mom saying to me, "Hold still, you big gaumin' thing!"

Though their children learn how to build a wood fire in the stove by the time they are five years old, most people believe stacking wood is an art best left to adults. Always pile wood barkside up if you don't cover the pile. They all know what a "night stick" is. They don't believe in yearly checkups, bathroom scales, dark maple syrup except in coffee or tea, classical music, fake Christmas trees, road maps, and hearing aids. Northern Vermonters never check the spare tire, their own weight, or the lawnmower blade. They don't bother to think about such things! Northern Vermonters like Mom believe the tune for "Beauty and the Beast" was stolen from "Moonlight in Vermont"! Northern Vermonters love everything raspberry, especially pie parties. They believe children should lick their plates after eating pancakes served with real Vermont maple syrup. Maple syrup costs real money and it must not be wasted! They believe in making friends with young people. Why be alone when you get old? Northern Vermonters never own more than four dogs or a dozen cats. They know never to plant anything in a month containing the letter "r", not even snow peas! They believe January and February should be spelled "Januairy and Februairy"! Northern Vermonters know not to trust any month ending in "ber" (brrrr!). They know school will never be closed for bad weather. Northern Vermonters are rugged. They are always prepared for the worst possible weather conditions, even if this means having the sense to "lay low" for a day, only going outside if they need to fetch one more armful of wood for the stove.

At a rest area on Interstate 89 I asked the attendant, "How are you on this fine sunny day?

He looked around and took his time to speak.

"Well, it's looking pretty good for the next ten minutes or so."

Mom loved summertime and for many years sold green and yellow beans and corn from her garden by the road. She would pick them fresh when people stopped in to buy vegetables. She always had a small chalkboard hanging under the mailbox showing her prices. One night soon after they were both in bed upstairs the chalkboard kept making a banging sound. Perhaps it was the wind moving it about.

Mom asked Dad, "Bernard, please go down and take the chalkboard down. I can't stand that noise all night."

But Dad was all tuckered out and couldn't crawl out of bed. So Mom went downstairs, unlocked the door, and marched out toward the mailbox. Suddenly Mom distinctly saw a Jersey bull standing at the edge of the driveway fighting the chalkboard with his horns. She dashed back into the house, trotted upstairs, and told Dad about the bull.

"You're seeing things!" he said.

Though half asleep, down the stairs Dad went to get the chalkboard. Out the door he flew. But just as he came near the mailbox, the neighbor's bull turned to see Dad. The bull let out a fierce bellow and stamped his feet at him. Dad walked backwards all the way to the back door. To the bull the sound of the chalkboard banging against the mailbox post told him another bull with horns had moved into the neighborhood! In another ten minutes the neighbor's bull figured out what was going on and meandered home.

In December 1945 Mom's Grandma Stone gave her a used steel frying pan for her wedding gift. She told Mom it was already 90 years old, yet Mom used it every day year after year, or rather "decade after decade". When we were visiting in 1997 Fran noticed she had a shiny new frying pan.

Being curious, Fran asked her, "Where's the frying pan that Gramma Stone gave you?"

Mom said, "Oh, I don't need that old thing anymore."

She had cooked with the same old frying pan for 52 years!

One time Stuey, Mikey, and I were telling Mom and everyone we didn't care that much about food. By this we meant we didn't care about consuming big fancy restaurant-style meals. But we loved Mom's meat and potato meals. And of course there was always a frosted cake on the shelf. Mom said she made a flat cake most every day for over thirty years!

Stuey told his mother very clearly which kind of cake he wanted: "I don't want chocolate cake with chocolate frosting. I don't want white cake with white frosting. And I don't want white cake with chocolate frosting. I want chocolate cake with white frosting!"

Besides eating much of it, Dad often ate all the crumbs he scraped from the cake pan.

So Mom said to everyone present: "They may not care much about food, but let them skip a meal and see what happens!"

She always cared that we ate well and she spoiled us with good basic meals and a variety of desserts, like her mother. We three brothers simply meant we didn't care for large servings of meat or fancy dishes. We loved Mom's good farmhouse cooking.

Cousin Alex joked, "I would have been six feet tall too if I had been raised on your mother's cooking!"

I have several tips for cooking and baking; not all originated with Mom. "Adjust" a cake mix by adding 2-3 tablespoons of flour, less than 2 tablespoons of sugar, and just a dash of baking powder. The cake comes out a little higher and somehow tastes almost homemade.

In 2010 I flew to Scotland to visit our oldest son. He was getting his Master's degree at the University of Glasgow. On the plane I met two women from Ireland who said they had stuffed their luggage with oodles of cake mixes. They said box cake mixes were unavailable in Ireland! I shared with them how Mom always "adjusted" a cake mix to make it her own. Four other little tricks are good for the palate. When making meat balls or meatloaf, stir in 2-3 tablespoons of spaghetti sauce per pound of ground meat.

Soon after people begin eating, you'll hear this: "Wow! Who made the meatloaf?"

You'll get fabulous reviews from this recipe too:

Cold or Warm Crab Salad
Chop together a red onion, 8 oz favorite olives, and 8 oz of crab meat. Eat as is, or
Add 1 ½ cups hot cooked pasta and shredded cheese.
If desired add mayonnaise to taste

When making a BLT try adding a fried egg and a slice of cheese! And be sure to warm the cookie pan in the oven for a few minutes as it is preheating. All the cookies will look the same!

The three of us older children grew up with the Round Oak kitchen stove. In the winter we would come in from outside and sit around it with our "stocking-feet" up on the opened door of the oven part of the stove. Then after many years of service our parents removed it from its place in the kitchen. Sadly it was parked in the woodshed for 15 years in favor of a gas kitchen stove. Now there was no more work of getting "stovewood" ready to burn every year.

But then the Round Oak stove came waltzing back into our farmhouse as big as life! Mom cooked and baked with their stove beginning in 1948. It kept the bedroom above the kitchen warm and toasty. A good cook stove is like a good friend. It invites you to stand near to it and makes you feel good on both the outside and the inside. The salesman was right. He told my parents the Round Oak stove would last a lifetime and always be there for them. And now it had its second life. Once again the big solid aluminum teakettle had its place to sit and make steam and hot water. It made Mom remember who gave it to her on her wedding day – her Grandmother Francena.

Mom often had two dozen warm cookies waiting for us after school.

"Take three cookies each."

When we begged for more, she would say, "Oh, eat them all. When they're gone, they're gone."

After we had devoured all the cookies, Mom would say, "Did you get one leg filled up?"

Mom especially loved babies. She loved her little baby brother Richard. She loved baby Marilyn, who Mom took care of at Lake Parker all summer when she was only thirteen. Babies always came in bunches for Mom. First came Kathy, Elizabeth, and me. We were all born at the cottage hospital in Barton. I was the "baby of the family" for eight years. Then came Stuey and Mikey, both born in Newport. When Mom got home from the hospital, she laid Stuey on her bed, stared at her newest baby up and down, and squeezed and kissed him all over.

She said, "After all this time, I can't believe I'm having another family!"

The next day Mom found a sponge ball, a baseball and a football in Stuey's crib.

"Philip, did you put these balls in the crib?"

I stood by the crib and smiled so hard that I couldn't answer her.

"You're going to have to wait quite a long time, my dear."

She gave me a big enough hug so that I knew she loved me and understood me. We all had a double delight when Mikey was born only two years later.

Mom always said the job she wanted when she passed on was to care for babies in heaven. Before she became a grandmother, Mom happily had "like-my-own grandchildren" from other families. Mom's cousin's grandchildren lived nearby and belonged to her too. Two more red-headed babies arrived in 1976 and 1978. Dad and Mom's nearest neighbors had two fine boys. A dear family friend adopted a special girl. Mom's nephew and his wife raised two boys and a girl on Perron Hill. All these children belonged to Mom!

So Mom was already living the life of a grandmother when, in the short span of three years and two months, her own four true-grandchildren were born! All covered in her love and kisses. Then after a lull in the action my brother Mikey's wife brought two more grandchildren into Mom's world. Mom wrote several pages about her third granddaughter in her baby book:

"I was puttering around the kitchen waiting to hear if Mommy would be having a baby boy or a baby girl, whichever, Grampa and Grandma would be very happy. Sort of thought of a girl, as I had in a dream. Little boys are so nice to, and Grandma should know. You dad was just adorable! We had notes and phone calls at first and a video of your other Grandma dressing you to leave the hospital. What a very nice baby and beautiful mother and happy dad you all were. Oh, how we love you, our dear little one! You have always known us, it seems. You knew us for certain – and didn't draw away. I knew instantly that you and I would like each other. In the house you came and looked all around. You were such a sweet little house guest and you were just unreal. Dad was so proud to see us together at last. I took you outside in my arms to the front lawn. A bird came with us and fluttered along. It was so nice to have you see the bird to say hello! What a wonderful long look you took – way out as far as the field went and to the trees beyond."

The birth dates of Mom's own five babies and six true-grandchildren covered exactly 60 years, from Kathy in 1947 her last granddaughter in 2007!

Mom loved her grandchildren. But no children were ever allowed in the "good living room" - only in the "television room". But she did not want anyone to pick up Gramma's toys after the children finished playing.

"Leave them right where they are. I want to look at them for a while after you go."

Mom loved to give good advice to a growing young mind.

"Go read the encyclopedias and the big dictionary. You can travel all around the world in those two books."

One evening when I was sixteen Mom and I traveled to the grocery store. I only had my learner's permit. It was almost dark. I could sense something was up!

Coming back from shopping in Barton, without warning Mom said to me, "Let's turn in here".

Soon we were at the straightaway of the half-mile-long Orleans County Fairgrounds horse track!

Handing me the keys, she said "Go in here and go around the track a few times."

I dreamed I was a race car driver but stayed in "the middle of the middle". Mom had no complaints. She seemed quietly excited to be on the track after dark, like something she always wanted to do, but had told no one.

A few years later after working only four days at the Ethan Allen furniture warehouse in Orleans, I was ready to quit. Handling cardboard cartons of all sizes made my hands very sore. All the muscles in my fingers ached.

"I load ten train cars every day. I'm never going back to that place".

"Oh yes you are! You are going to work at the mill this summer."

Mom was full of "advice" and was always sure she was right. I grumpily conceded to her "wisdom".

Mom loved her northern Vermont summers. And she especially loved the Lord's Prayer, often speaking about it in detail. She lived and died by this prayer and found Jesus to be a wonderful Savior.

Bernard with his father's milk truck 1940

4

Bernie

A poem for **Bernie** on his 96th birthday, January 5th, 2018:

January Fifth, 1922

January Fifth, Nineteen Twenty-Two
Do you know if the whole sky was blue?
Surely it was so terribly cold!
But it wasn't really so long ago,
If you are ninety six years old.

Your father was happy for another boy;
Your mother called you her winter joy.
They loved to put you in their bed,
A tiny one with precious face and head.
You blinked and winked and slept;
Around her thumb your wide hand curled,
Your golden hair her fingers twirled.

Their love for you could never end,
Like the love of Jesus our true Friend
So be happy to be loved by God.
You are His child now, don't think it odd.
Life begins with Him, and never ends.

Like his sister Harriet, Bernard died in his sleep in 2018 only 12 days after his birthday. Nine decades earlier little Bernard and his brother Edwin asked their mother and father if they could go to church with their grandfather. They skipped happily down the road for half a mile past the

Andersonville Cemetery to "Cooper Johnnie's Corner" near the exact place where Glover, Craftsbury, Albany, and Greensboro all meet. They waited patiently for their grandfather to arrive from the Rowell farm in South Albany. And now Bernard has gone from the earth to be with his grandfather John forever. You see, they believed the promise of Jesus, Who said, "Because I live, you shall live also." (John 14:19) And so now they live with Jesus! When Jesus returns all creation will be restored, even our bodies, every one.

Bernie

Dad hadn't shifted gears for several minutes. I slid over to the middle of the seat and placed my right hand on the shifting knob. Then I lowered my head and put my right ear on the back of my hand. Dad taught all three of us how to do this. The sound of the gears told me we were making good time. Dad had both forearms resting on the big steering wheel. He looked happy.

"Not many people live on this road."

"Yuh."

"Why is this such a wide road when no one lives here?" I asked.

"It's for the politicians."

"Whatta y' mean?"

"So they can get t' Montpelier faster" Dad said, as we skirted around Greensboro Bend.

A little later I said to Dad, "My toes are freezing."

About a mile up the road he stopped the truck.

"Walk up t' the next farm."

I hopped out and watched Dad's truck go up the hill in the dark. It seemed funny to be standing on the edge of the road in wool socks and barn boots. As I turned to walk I realized how cold and tingly my feet and toes were. My ankles were not happy either. As the snow on the road crunched and creaked under my soles, I finally got some relief. But now my coat didn't seem warm enough.

I was glad Dad had been driving truck since he was sixteen years old. He had learned a lot of tricks to beat the cold. I thought about how Dad brought the battery into the house at the end of the day, setting it on the hot air register where it warmed up all night. Then the battery had more strength than if it was left outside under the hood.

It wasn't quite morning yet and I could just hear the truck backing into the farmyard. I wished I could put my ear again on the knob of the shift lever and hear the whine of the gears in reverse. Only the ends of my toes were cold now. When I walked by the truck I could hear the farmer asking my dad something.

I heard him say "Yuh . . . tomorrow."

"This your boy?"

"Yuh, this is Phil."

"Well, good. Well, thanks Bernie."

My Aunt Harriet called her brother "Bun", but I had never heard anyone call him "Bernie".

One time Mom said "Who would name their son "Bernie Ernie Urie?" But Dad told us he had a distant cousin from Barnet or Passumpsic also named "Bernard".

After he loaded the nine milk cans, Dad filled out the receipt and stuck it on the nail on the wall. Then he turned for a moment to look in at the cows. I couldn't take my eyes off him. Something had happened.

I said to myself, "It must have to do with what they were talking about."

I thought about the time a few months earlier when Dad seemed so disappointed with me. I just couldn't help it. He wanted me to learn to milk cows, but I was too afraid of them. One of them kicked at me and I cried. Seeing the way Dad looked in at these cows and not knowing what the farmer said to him made me nervous. I felt like I was less than a boy. Suddenly I knew I was just a "little boy".

To make matters worse I remembered playing in the big field out front that summer for days and days while Dad used an iron bar, a pointed shovel, and his own hands to remove big rocks and level out hummocks in the big field out front. He placed the rocks on a stone boat and the workhorses, "White Sugar" and "Brown Sugar" drew them to the brook, where Dad dumped them along the edge. The idea of being almost no help to him hurt me. I was in his way.

I watched Dad wrap and tighten the two logging chains around the back doors of the truck. We got in and went down the road.

"Dad, why did you sell our cows?" I asked.

"I didn' wanna borrow the money for a bulk tank."

I thought how maybe he would have borrowed the money if only I was older. Uncle Johnnie's first child was a son. I felt bad for my father; and I felt sick inside.

Dad pulled out a bag from under the seat.

"Wanna' chocolate?"

"Thanks" I replied.

They were all chocolate-covered cherries. Dad ate two, but I ate three of them.

After several minutes Dad said, "Someday we can get a used milk tank."

I wondered if we would ever farm again.

Not long after this, Dad sold his truck and went to driving a brand new bulk tank truck for *Buttrick*'s creamery in East Montpelier. It was so fancy compared to his plain old can truck! It was maroon and cream-colored with two rear axles and air brakes. Dad seemed very proud of it, though somehow he looked smaller when he climbed up into the cab. I only listened to the sound of the gears one more time. It seemed like something only little boys do.

Over the next eight years Dad milked cows for several farmers morning and evening six days a week. This was in addition to his job driving truck. He would arise each morning at 3:30 and get home at ten minutes before eight in the evening!

Elizabeth and I liked to go with Dad and stop at Smith's General Store in Greensboro Bend. Dad had an account there to purchase gasoline for the car on his round trips to and from East Montpelier and West Glover. He would often buy 25-cent ice cream cones for the three of us. We picked our flavor and Mr. Smith put two scoops on the cone. Then just as we were about to take it from him, he always let it roll upside down in his fingers! We didn't mind, because an upside-down ice cream cone was always more fun than a regular upright one.

Even after we installed a used bulk tank and began farming again, Dad found part-time work. He worked for his cousin Jean and her husband. They owned the second largest farm in Glover. After breakfast Dad would go up and back the big manure spreader into the heifer barn and the shoveling would begin. With Dad on one side and Alden on the other, they both scooped and lifted and threw the manure up into the spreader. Glancing at each other, they tried not to sling the wettest manure into the spreader at the same time! They were both steady and aggressive workers.

Alden often did our haying after their hay was put up for the winter. One day Alden was shaking out some wet clumps of raked hay by hand. The weather had been overcast for days but no rain had fallen.

"How do you like the weather lately?" I asked.

"It seems there's quite a lot of sameness to it", Alden replied.

One time in Pennsylvania, I used the word "oodles" in the course of normal conversation.

"Where did you get that word?"

I thought of Alden's word "sameness" a perfectly properly used word.

Their son Kenneth wrote to us about our Dad after his funeral.

"Bernard was a big help to us when we purchased the farm from my parents, as my dad's health was failing. He would do whatever we asked of him, cutting wood, sugaring, haying; he even helped me lay sheetrock. . . When we were haying, I would tell the teenage boys to sit in the shade between loads of hay, Bernard would go and split wood. I remember a day in April when he and I were pulling down buckets. Our last load came from the valley below the sugarhouse. We unloaded the last load of buckets at the sugarhouse and breathed a sigh of relief that we were done. We looked down into the valley and saw one lone bucket still hanging on a tree. Before I could say anything, your dad said, 'that was my side of the road', and took off at what I can only describe as a gallop to get that bucket!"

My cousin Debbi, once said: "I remember your dad as a man who was always working."

He never had a real extended vacation; hardly even one day here and there. Dad worked for others until he was 85.

At the nursing home Dad said to me, "I can't believe how much work I put out in my life."

Here is a list of the places he worked, most all in the Northeast Kingdom of Vermont:

His father Ernest's milk can pick-up business named "Speedwell Dairy" in Andersonville: At 16 years old Bernard was known as a truck driver who never got stuck. Vanasse Builders told me John Jr. was the same type of driver. He could drive his loaded cement truck onto a muddy construction site, unload, and leave with no problem. Ofttimes other drivers had to be pulled out.

Bernard was part of his father's logging and sawdust business. He also milked cows for his father by hand as a youngster.

Helped his father and brother John build the Kancamagus Highway in central New Hampshire

His own trucking business, plowing snow, hauling logs, and delivering sawdust for bedding cows: I remember leaving the sawdust truck at the "Bobbin Mill" in South Barton. This business made wooden bobbins (spools) for the woolen yarn industry. After three days Dad would pick up the sawdust and deliver it to a farmer for $12.

Plowed snow for the Town of Glover: Dad owned a Ford Coleman 4X4 truck complete with a side plow. That truck sat unused for over 30 years until Bob got it running and bought it for hauling firewood for himself. Most houses in Northern Vermont burn multiple cords of wood each winter and need more than just a pickup truck for hauling wood.

Hauled milk into *United Farmers*, Barton and *Findyson*, Lyndonville: H. P. Hood also had a creamery in Barton and in scores of towns and villages in northern New England. Hood fieldperson Norm Lucier wanted me to replace him in the Enosburg area. *Hershey Foods* had over 2,000 farms in Pennsylvania.

Drove milk truck for Joe, Plainfield into *Buttricks*, East Montpelier: Dad never purchased a bulk tank truck after the state banned the use of galvanized steel milk cans.

Hauled bulk milk truck direct for *Buttricks* with their own truck: The milk plant processed milk from about 25 farms in Woodbury, Calais, and Marshfield. *Buttricks* claim to fame was that they delivered milk to all the Kennedys and others around Boston and Cape Cod.

Tested producer milk samples for butterfat at *Buttricks*: Farmers (producers) were paid on the basis of butterfat percentage formula. Holstein milk averages 3 ½ percent. Today farmers are paid on the basis of both the fat and the protein percentage. Holstein milk averages 4 ½ percent protein.

Hauled milk for Earl into *Kraft Cheese*, So. Troy, Vermont: Earl was raised in South Barton (village of Kimball), attending a one-room schoolhouse, just like Dad attended the one-room Andersonville school through 8th grade. Dad highly respected Earl.

Milked a large herd of over 50 cows in Plainfield: there was no "dumping station". Dad dumped the milk into pails and carried them to the milk room. But he didn't have to feed the cows or "wash the pails", just milk the cows.

Milked cows both morning and night for a cattle dealer in East Montpelier before and after hauling milk into *Buttricks*: Bob was a cattle dealer, as was his brother Ben in St. Albans.

Dad helped stretch thousands of feet of barbed wire and woven fence for a beef cow herd on Burton Hill.

Worked in field and woods for dairy farmer Urban, Irasburg

Worked for Leo and Mike, Barton trimming Christmas trees together on the hot days of July and August

Produced maple syrup at a brand new sugar house on Roaring Brook Road, which burned only fuel oil, not wood. Dad would simply "throw the switch" if the fire was too hot!

The maple orchard was called a "cold orchard" because it faced the north. When sap had been running for a week in most orchards, this one was just starting to run.

Worked as "mud man" for a good mason: This was one of Dad's favorite jobs. He really enjoyed Nelson's comments along the way. Dad worked with Nelson building chimneys, including two fine chimneys for Dad and Mom's own farmhouse. One time Nelson measured the straightness and levelness ("the plumb") on the chimney they were working on.

He smiled and declared, "Not too bad for people who are used to not having the very best!"

Trimmed hedges and did yard work for a widow in Barton: Her husband, Emory Hebard, owned a grocery store in Glover village and went on to become the Vermont State Treasurer. When I first started working for the state, my paper checks were stamped with his signature.

Tended the outside Community Ice Rink, Barton: Mom wrote to us on February 3, 1989, "Dad is at the rink today. It is growing colder. He needs something to keep busy at and this is fine."

Worked in farm, fields, and woods and sugared for his cousin Jean in West Glover: About her husband Alden, Dad said: "He taught me how to work."

Dad's own father Ernest had taught his three boys to crawl out of bed at the sound of his stick banging on the stovepipe. He kept tapping until they came down the stairs! A man I worked with told me how, when he was only five years old, his dad came into the bedroom at 5 o'clock one morning, removed his belt, and folded it in half.

"Now, starting today you'll get up at 5 o'clock and go to the barn with me every morning."

When Dad was 80 and Mom was 75 they climbed Lone Tree Hill with us. Years later Dad asked Gary (whom he had worked for) about planting 10-12 maple trees along their shared property line.

Gary joked: "Go ahead Bernard. You'll be tapping them long after I'm gone."

Worked on the farm and trimmed Christmas trees nearby with Bill in Barton and Montgomery.

Another farm where Dad worked was the LeClerec farm in Barton. The farmstead looks to the southeast near the top of Burton Hill. Dad was working with Nelson to help put firewood away for the winter. Using a manure spreader as a "wagon", they hauled an overflowing load of wood chunks cut from fallen trees at the edge of a pasture. After unloading the spreader Nelson noticed a chunk of wood lying in the driveway.

Looking at it, he said to Dad: "Musta been a good draft in the chimney when Bud threw that one in the furnace, Bernie!"

My Dad was "Overseer of the Poor" for several years in the 1950's. One cold winter day Dad and Mom travelled to Miller's home in Andersonville with a box of canned goods and a few fresh groceries. We parked along the road because they were "snowed in". Elizabeth and I trudged our way behind our parents, following in their footsteps in the snow. Using one of our sleds, Liz and I dragged the box of food up to the front door. A tall boney old man, all stooped over, came to the door. He didn't want to take the food, but he did want us to come in and visit. A few minutes later his wife came trudging through the snow carrying a pail of water in each hand. They had no running water! There were holes in the walls of the house in several places. The whole house smelled of firewood which wasn't dry enough to burn properly.

Mrs. Miller walked over towards an old piano, saying, "There's a bag a' candy hee-a somewhey-a".

She found it and Liz and I politely each took one. They didn't look very fresh. We placed them in our pockets.

Before we left the Millers thanked us for the food, but seemed so ill-at-ease. They both had tears in their eyes.

We have a photograph of our little brother Stuey imitating Miller. He is standing stooped over with his hands on his legs just above the knees.

"Look! I'm Mr. Miller!"

One time Uncle Johnnie shared with us all how he and "Bun" used to go to Daniels Pond in February to cut ice to store for use in the summer. The pond is almost a half-mile long. I asked them where on the pond they cut the ice. Dad indicated it was at the far end of the pond near the outlet to the southeast.

When I asked him how they got there with the horses, he said, "Straight across the pond on the ice. We didn't use the road"

"That sounds dangerous! Why did you ever do such a thing?"

With great firmness, yet with a grin, Johnnie chimed in: "Because our father said so!"

"Did you go back on the road?"

"We followed our tracks back across the pond, just like our father told us to do."

One fall day Dad asked me to take his chainsaw and cut off a big branch that was growing too close to the barn. He said he would cut it up for firewood. It seemed dangerous to me because I would have to hold the chainsaw over my head. I suggested having Uncle Johnnie take a look at it. Johnnie came right down in his Subaru with his chainsaw in the back. He nonchalantly picked up the saw by the end of the bar instead of the handle.

"Where is this tree?"

Dad led the way out behind the barn, us in tow. Johnnie looked at the big branch from several vantage points.

"Do you need this wood now?

"No."

"Then let's let it grow one more year."

The following spring I noticed the branch had been cut off.

"Who cut down the big branch?"

"I did."

"How did you do it?"

"I set up a ladder on the back side, climbed up above it, and cut it off."

Dad was clearly happy with himself.

In 1979 Dad put an addition on the barn for the heifers and dry cows. The first thing to do was dig for a foundation. Stuey was away in the Navy, but Mikey was there to help dig.

After a few hours of digging over three feet down, Mikey said, "Couldn't we hire someone with a backhoe to dig?'

"No. This way we'll know how it got here."

In 1988 Dad also built many more barns – children's play barns, that is. Mom collected little plastic farm animals and made grain bags for the animals too. Children played for hours with their "Grampa Barns". Dad also made a "Bucking Bronco Horse", stilts, and a walking puppet-on-a-string bird. He said he got his ideas from his grandfather John.

My youngest brother always said his voice and way of speaking sounded more like Dad than he himself did! Dad had a pleasant fatherly voice.

A man on a television show said to his wife, "I just looked in the mirror and saw my father!"

When I told these things to a man at work, he answered quickly, "That's nothing. When I look in the mirror I see my grandfather!"

It is a thing to treasure to sound or to look like someone you care for.

When Mom went to the nursing home, Dad decided to sleep in the bedroom above the kitchen stove just like he did when he and his two brothers were young children in Andersonville.

"I know why you are sleeping there. It's because it reminds you of your childhood."

"I've been waiting fifty years to sleep up there", Dad said with a smile.

Uncle Johnnie said they slept three boys in the bed when they were small. Think of how cold it was by morning! But at least it was the warmest room in the farmhouse. I asked Johnnie who slept in the middle.

"Most of the time your father ended up in the middle. Edwin and I kept him warm."

After Uncle Johnnie passed on, his wife Aunt Pearl kept right on burning wood in the furnace in the basement all winter. Her furnace could also burn fuel oil. Her children began to worry that Pearl might fall down the stairs. They decided to wire the furnace door shut so she wouldn't have to go down the stairs ever again.

"Wood heat is the best heat and that's what I'm going to have! Go ahead and wire it shut. I'll just hire someone to unwire it so I can burn wood in my furnace."

Fondly I remember Mom making cake, raised doughnuts, and beautiful rolls with her Round Oak kitchen stove. She even let us each eat several doughnut dough holes. Mom always wanted me to be a heartier eater, but I imprinted somewhat on Dad's light eating habits.

Mom quizzed me: "You probably make only three-and-a-half tablespoons of oatmeal, just like your father."

"Yup", I said proudly.

But in one area of eating I surpassed Dad.

I once told Nate: "For every chocolate candy Dad can eat, I can eat five!"

One time my Aunt Harriet happened to ride her bicycle down to West Glover when Fran and I were visiting Mom and Dad. She came up to me and hugged me.

She seemed surprised and said to me, "My, my! I guess when you grab a Urie you hafta' get used to grabbin' bone!"

A young framer once said to me, "You're not thin, Phil. You're just lean."

I can never remember Dad going for "second helpings" unless it was for a desert once in a while. Dad was always a "careful eater" and a good example to us of how to not over-eat.

One time at church, Dad said his grandfather John noticed little Bernard tapping on the pew and distracting a few people. His Grampa placed his big hand completely over his little hand.

"Did he frighten you?"

"No, no, no. I knew he loved me."

Bernard loved his family. Following are samples of Dad's dear letters written to me:

Dearest Philip and all the family too, I am looking at the picture card of Calvin taken on his first birthday. What a good group we all were then and are still the same loving kind of people. Your mother and I really do think of you all every day and am sure you do also. Mother's name is on the name plate (return address sticker) *and I love her so. It is hard for us to stop and think where you are at this very moment. God is well and Christ be with us all. Dad*

Hello Philip, Thank you for the two bits. Calvin's letter came today the same as yours did. You gave me a pair of felt liners, but I don't think they will fit my boots. It sure is cold up here. They are feeding the birds. Jesus is helping me every day just like a friend. I am going to put this note in the envelope which you sent, trusting Him always. Love, Dad

Dear Philip, Yes, the leaves sure have given us a wonderful season, so beautiful. Now it is a different time here, up north anyway. The white season will be on us soon, and we shall have to take whatever comes. It's raining right now and the chickadees are getting a free bath out on the boardwalk. I just ordered my cornflakes for supper. Will soon eat them. Did you know Liz has a tame squirrel about the house! Thank you for the money. With love in Jesus Name, with hugs and kisses. Dad

Dear Philip, What a silly card that I sent to Nate the other day. You know how I am, just pick up any card and send it! I do hope you will appreciate it never the less. Whenever I see your mother, she always asks to go home with me. Of course it is impossible because it takes two nurses just to take her out of bed. It really breaks my heart. I have oatmeal and toast for breakfast, hamburg for dinner, and cereal for supper most of the time. My sister Harriet brings down some goodies. She sent a whole plateful Thanksgiving Day. It was so good. Hurricane Sandy must have made your house rattle around! Will close now with love. Dad

The Magic of the Farm

Blessed are you who weep now, for you shall laugh.
(Luke 6:21)

5

Village Life

My childhood home was a working dairy farm in West Glover village. My parents owned 25 Registered Holstein milking cows plus young stock on the road to Barton, called "Roaring Brook Road". Dad used to put a salt block on a short post in the pasture for the cows to lick. The cow's tongue's left depressions in the salt block. The three of us liked to lick it too! Our milk was shipped to a Kraft milk plant where it was made into Velveeta cheese. Dad always supplemented his income by working part-time for other farmers. Like most wives back then, Mom worked outside the home infrequently.

From the time I started attending the University of Vermont in 1970 until I graduated in 1974, I was homesick every day for what I describe in *Village Life*. I yearned to return to West Glover. Our farm in the village was far from heaven, but like many my age the farm was the place to be and the center of the world for me for a long time.

I was home again until one July day in 1974 when my mother came running out of the house in her bath robe and slippers. She stood inches away from me and was a flurry of words.

"There's not enough here for you and your father. You're out of here today!"

I said nothing. By the next day I felt like I had been run out of the place I grew up in, like some kind of a bear cub spat off by Mama Bear just before winter set in! My thoughts travelled to many of my class mates who had "stayed on" the family farm. But what my mother said was true and it was time to go. My two younger brothers would take over for me anyway.

The next day I found myself four hours away, far from the village life, working as a herdsman at the other end of Vermont.

Village Life

West Glover village is very quaint and surrounded by larger than average fields, with cows roaming the fields from July through October between the end of haying season and the beginning of "the cold", usually a day or two after Halloween!

Dairy cow eugenics were practiced nationwide once artificial insemination began in the 1950's. At first only veterinarians were allowed to inseminate cows with fresh semen. Soon it was discovered frozen semen could be thawed without loss of viability. Also only a small amount of semen was needed, and so individual bulls sired thousands of daughters. Bulls were pictured in a "sire catalog" and were a true retail business. Many men and farmers became "artificial breeders", able to make a modest living, especially in areas where cow density (population) was high. Franklin County still has the highest cow population in New England. In 1975 there were 600 farms in Franklin County. The largest dairy cow county in the United States was Marathon County, Wisconsin with 3,000 farms! Bulls had to be "proven" to remain an "offering" of the bull stud. First bulls with the best pedigrees were used all over the country for several months. Then they stopped using him in case the bull turned out to be no good. It took five years for the bull to come out with his first "proof". Only "the cream of the crop" bulls were kept "in service". At least 95% of the bulls were rejected! Calves looked more and more promising and grew into cows which far excelled the size and quality of the previous generation.

Dad loved to quote what his neighbor Leonard said about good cows that were getting bigger and better in every way, saying "Quality, quality, quality."

Today genetic/chromosomal science has sped up the whole process, and any personal knack in the area of profitable mating is virtually immaterial. "Cow men" have almost disappeared from the farming scene. It's all about using cold science to rid cows of bad genes. Eventually all cows will be very much the same. Matching cows to the right bulls was part of the magic of the farm. Now the day of small farms has fizzled out, all in only one generation of farming. Dairying has become the large-scale business of a decreasing number of farmers who own the majority of the best gene pool. There are now only two farms operating in the whole town of Glover.

Lone Tree Hill stands above the fields to the southeast of West Glover village. Until the 1960's a beautiful American elm tree topped off this hill. Several elm trees also lined the village street between the school and the store. Merle Young Sr. was the self-appointed "mayor" of the village.

Merle Young Jr. once said, "Everyone wants to live in West Glover."

A modified tale originally told long ago by University of Vermont professor Francis Colburn describes how special NEK people are:

"You've been at Parker Pond and shopped here all summer, Pete."

"Yes, Judson. I wanted to stay long enough to get to know some true Vermonters here in the village."

"So what do you think?"

"I think Vermont sure has some odd people."

"Ey-yut. One thing about it though - they all go back come fall."

Another Colburn story tells of two out-of-staters who got lost in the maze of dirt roads in the Northeast Kingdom. But soon they came upon an old farmer scything grass on the side of the road.

"Can you tell us how to get to Boston? There were no signs in the village."

"Well, if I were goin' t' Boston, I wouldn't start from hee-a."

Northeast Kingdomers have a certain "truth-speak" combined with seasoned practicality, which is styled "dry humor".

One time a reporter challenged another reporter about the Vermonter and President Calvin Coolidge.

"He is so reticent. I bet you can't get him to say three words."

"Mr. President, my friend says I can't get you to say three words."

After a short pause, Mr. Coolidge responded, "You lose!"

My seminary professor Spear asked me if it was true that Vermonters have little to say.

"In the winter."

"What do you mean?"

"They don't want their teeth to freeze!"

He thought I was joking!

Vermonters' speech is sparse because of the cold. Short sentences suffice in the winter.

For instance, I once heard a man succinctly say about his boss: "He don't tell nobody nothin'. " This statement may be a triple negative, but it got the idea across with just a few words!

Northern Vermonters keep their lips shut for so long in the cold weather that it becomes a year-round habit to offer only short sentences in any conversation. But get them inside the house and they talk and talk, especially if they sit near the woodstove!

Truly the main feature of the Northern Vermont is "the cold". The cold is a sure thing and a very present reality. It must be dealt with at all costs. Only the two months of July and August are reliably comfortable. Only then can you safely forget there is such a thing as "the cold".

The land here is still peppered with small but now mostly empty barns. One hundred years ago a visitor from Europe described America as "a land of steeples". But it is also a land of barns. Eric Sloane wrote about and painted America's small farms. In another 100 years the barns will most all be gone. In Vermont in the 1980's a very conservative political party was founded by John McClaughry, a state representative from Kirby. The new party was called "Small is Beautiful". This party sought to conserve small businesses, but especially small dairy farms. Quebec has preserved their small farms for decades with government financial support.

Bernie Sanders' urban roots meant he had little interest in farmers or in farming. But the farming community adopted Bernie into their ranks because he could deliver a good speech to serve their needs. Farmers liked their way of life and at least Bernie aligned with their independent spirit.

Vermont's small farms were built up in value by much free labor. Amish and Mennonite families do the same thing on their farms. Most children develop their farming and building skills and make each other rich with free family labor and "barn raisings" for all. A lazy child is titled a "hooftee" or a "heeva-hava". The whole rest of the world who the Amish call "the English", also get free labor from their sons; their children are not expected to stay on the farm after high school like the Amish do. They become "hired hands", staying only until they get their feet under themselves. In the Northeast Kingdom many young men ended up working at mills and furniture plants for quite low wages. But the advantage is that they don't have to live on the back end of a cow all their lives! Recently wages at "the mills" have increased significantly.

The biggest change in West Glover village is "Parker Pie", a restaurant named after Lake Parker that used to specialize in pizza. Now they specialize in beer, or shall we say "beers". You may go to the restaurant today and see the old entrance door with deep dog scratches embedded in it. The scratch marks look exactly like they did in the 1950's when we went up to the village to buy penny candy. Soda was 7 cents and candy bars were 5 cents. The "West Glover Store" was operated by the two Wanamaker sisters and their brother back then.

We swam in Lake Parker at the state boating access often because it was only a five-minute bike ride away. When we were growing up, there were 13 houses in West Glover village. Today there are still only 15 houses, but many more on the outskirts. Two of the three barns are still parts of a large dairy farm in the village except for our barn. It has had no dairy cattle since the USDA Herd Buyout in 1986 and now the whole farmstead has been sold in three pieces.

West Glover had a two-room schoolhouse in the exact middle of the village. One teacher taught the first four grades downstairs and another taught four grades upstairs. Once when I was in the First Grade our teacher, Mrs. Davio, sent my cousin Wayne and me up the street to her house to watch the baseball World Series with her husband! The only station most Northeast Kingdomers could "pick up" on the antenna on the roof was *CHL-TV* in Sherbrooke, Quebec. So we turned the volume off on the French-speaking television and played the American radio to go with the silenced television!

On the coldest winter days we all moved our school desks to form a huge semi-circle around the kerosene stove.

Mrs. Davio never spanked anyone. Instead she would stare at you and growl with a smile on her face. When things really got out of hand she spread a rope loosely around a few of our chairs or had us sit under the teacher's desk for punishment. We all loved her.

Community bingo parties were held upstairs in the evening to raise money for the West Glover Village School.

The school was torn down a few years after it was closed in 1963 in favor of an expanded "Glover Community Elementary School". There used to be five neighborhood schools in the Town of Glover. The village children who walked to school now walked to meet the school bus, waiting outdoors at the West Glover Congregational Church. It was very cold standing on the steps, but at least it faced the south.

Other buildings included the former Meadow Brook Creamery, first owned by the Aldrich family and then the Stevens family, a house that originally was a hotel in stage coach days, and a very small post office and library. West Glover was originally known as "Boardman's Hollow". A dungaree factory and a lumber mill were here in the late 1800's, located just above the bridge, utilizing water power from a log dam. This long-abandoned log dam created great fishing opportunities for many village children. One time I caught a huge turtle! I cut my line and let it go! David Squires caught a 22-inch brown trout under the old logs, and appeared with Tony Adams on *WCAX-TV*, Channel 3 in Burlington describing his prize-winning fish.

Almost every house and barn including the creamery, got their water from a spring in the Town of Barton three-fourths of a mile away at Meadow Brook Farm. In 1958 the supply line was replaced by several people, including our farmstead's three-quarter- inch plastic pipe.

My parents kept two woodstoves going all winter – a cook stove and an ornate Ben Franklin stove between the dining room and the living room. Neither stove had a window. So Dad and Mom never sat and enjoyed the flames like we do with our *Regency* wood stove. To know, as long as you have dry wood, you have heat and can heat water in a pot if necessary, is very comforting. Watching the flames is an "extra". The heat is the main thing.

For many years we got together with two sets of cousins for "sugar-on-snow parties". We often called them simply "sugar parties" complete with hot maple syrup spread like a golden ribbon on snow packed in a pan ("sugar on snow"), both homemade cake and raised doughnuts dipped in warm syrup, deviled eggs, whole pickled beets, bread-and-butter pickles, *Cabot* brand cheddar cheese, chocolate milk, and coffee sweetened with maple syrup for the adults.

Northern Vermonters relish first-run or "fancy" light syrup, which I explained to an ignorant flatlander at the *C&C Market* in Barton. It was not that he was stupid. He just thought darker syrup was better. That's what some Vermonters say, so they get to keep "the good stuff" for themselves! Dark syrup is always somewhat disappointing to the palate. Following is my favorite pudding desert using maple syrup, given its name by Selah at church:

Banana Great!

Make or purchase a graham cracker pie crust

Combine and pour in a well-mashed small banana and a tablespoon of maple syrup

Pour on up to 2 cups of cooked vanilla pudding, once it is cooled somewhat.

Refrigerate and then serve with whipped topping, if desired

We buy our maple syrup from either my cousin's Andersonville Maple south of Beach Hill or Taylor's family business in West Glover. We are very particular! Taylors tap several thousand trees all around a hill 1860 feet high on the Albany-Glover boundary. High elevation and good exposure to the south are key factors in making good tasting maple syrup along with quantity. The view from Taylor's farmhouse looking to the northeast into the Quebec Laurentians is astonishing. Doreen Lyon recently produced a beautiful book about her experience as the 1970 Vermont Maple Queen. But the bulk of the book is a batch of fabulous maple recipes. There are 83 pages with 109 recipes, entitled *Recipes from a Maple Queen*. Beautiful photos are included on every page.

Once I built a bridge across Roaring Brook by felling two large poplar trees all by myself with only an axe. They practically fell into place about five feet above Roaring Brook. I found some boards in the shed to finish the four-foot-wide bridge. We used it for snowmobiling between us and the village behind the old creamery. Now we could avoid the roads and the bridge in the hollow in the center of the village. My bridge lasted three or four years until high water washed it away and broke it up in pieces one night. Although Roaring Brook "roars", there is not a single waterfall along its 2 ½ mile journey from Lake Parker to the Barton River, though there is an elevation drop of over 500 feet.

Looking northwest Jay Peak (elevation 3,871 feet) is in full view from the top of my parent's farmland on Bean Hill in West Glover village. From our home in Swanton we could see the Adirondack Mountains in New York to the west, but nothing of the Green Mountains of Vermont to the east. But one day when I was on the roof cleaning the chimney of our home, there stood the faraway top of Jay Peak. My field of view to the northeast followed the lowland of the Mississquoi River Valley eastward and on toward only one mountaintop over 40 miles away - what I always thought of as the insignificant back side of Jay Peak. It is a two-hour drive from Swanton to West Glover, with Jay Peak posing like a lighthouse shining halfway between two worlds which I know well.

6

From Farm to Farm

From Farm to Farm are vignettes of my experiences as a Federal and a State Animal Health Inspector in both northern Vermont (1977-1992) and eastern Pennsylvania (1995-2015).

The best dairy farming region in Vermont is the Champlain Valley, including Franklin and Addison counties, the two counties north and south of the big city of Burlington. New Hampshire boasts of having the rich soil of the Connecticut River Valley. L'Estrie (the Eastern Townships) produce the best crops for Quebec farmers. The "Farm Coast" is Rhode Island's little breadbasket beside the ocean. The Garden of England is the Weald of Kent. Pennsylvania's Lancaster County, west of Philadelphia, is the center of the state's best crop and plant growing region. Pennsylvania farmland there has sold for $100,000 per acre! A stone or ledge is rarely found in its productive red soil. One time a farmer's eighteen-year-old son told me he wanted to relocate and farm in "my neck of the woods" four counties north in Bradford County, Pennsylvania where the deer hunting is better.

"Do you know what rocks are?" I quizzed him.

The farmer's son just stared at me in deep thought.

From Farm to Farm

On my first full-time job away from West Glover, I found myself four hours away at the southern end of Vermont. On the second morning I was left all alone milking a herd of 60 Jerseys, which were not my favorite kind of cow. It was all new to me to be milking in a "milking parlor" standing below the cows in "the pit". One cow had a bad habit of kicking while being washed for milking. So I climbed up and put "the cranks" over her back just in front of her hip bones. This is a contraption shaped like a huge horseshoe which can be cranked just tight enough to keep a cow from kicking. By the time all the cows were milked on her side of the parlor I forgot I had "cranked" her. I opened the gate and out went the cows. That's when this "kicker cow" waddled out and fell into the pit, the place meant only for humans. There she was lying at my feet! I hastily removed the crank and she jumped to her feet. She walked up the stairs! She didn't seem to mind the experience and looked at me forgivingly. Perhaps she had fallen into the pit before!

One morning this farmer told me he would be away all day. He told me to bale up the hay on the "seven-acre piece". What he didn't know is that I had never baled hay! I was a herdsman, not a field hand! But after lunch I went down U. S. Route 7 pulling the baler and the wagon all by myself. After I made only a dozen bales of hay the twine broke on one side. I had never fixed a broken baler string but was successful in following the pattern of the good string. When I continued baling, the first bale came out with a piece of baler twine 10 feet long hanging from it! I was happy to have no more problems.

I worked at one more farm before I gave up on being a herdsman. I really enjoyed working with Registered Holsteins for Ray in Colchester.

Arthur Conan Doyle said: "The work is its own reward."

But I found it so difficult to have only half a day off every two weeks!

It was here that I saw a fluffy Eastern wolf roaming the fields next to swampland around Mallett's Creek north of Nourse's Corner. In one of the pastures there is a fantastic wide waterfall about eight feet high.

In 1975 I moved to St. Albans hoping to find some kind of work instead of milking cows. Soon four local men were hired by Veterinary Services, a part of the USDA. I was one of them. It was a good thing, for I only had $10 to my name. I ended up living on bread and water (yes, literally) for the next 10 days! On a Friday everyone except me got their USDA check for two weeks' work. So I had a robust hope to see my check the next day. Instead I waited the whole next week and saw no check, not even on the second Saturday. I walked over to the nearby *Grand Union* store and bought another loaf of *Roman Meal* bread. Monday rolled around with "visions of sugar plums dancing in my head"! But it was Tuesday when my first check finally arrived, eleven days after everyone else had been paid. But when I looked at my check, my name was spelled wrong! I told my plight to the bank teller.

"We'll cash it and say nothing".

Only three days later I received my second check.

The four of us were charged with blood testing every cow in Franklin County in northwestern Vermont for brucellosis, a bacterial disease of cattle which causes late-term abortions. In humans this disease is called "undulant fever" and has been described as the worst disease known to man. In 1975 at the age of 22 I was trained by a federal veterinarian from Illinois, who was head of the "task force" of many state and federal employees.

While being trained at the Richard Powers farm in Richford, the task force veterinarian said to me, "You should become a veterinarian."

I never did.

It took the four of us 20 months beginning in October of 1975 through May of 1977 to test 50,000 cows in the 600 herds in Franklin County. The four of us also each managed quarantined herds. To this day Franklin County, Vermont has more cows and makes more maple syrup than any place in all of New England. Many roads have no houses, except for the farmhouses. The 1970's were good years for the dairy farmer. Jimmy Carter raised the price of milk to over $17 per hundredweight.

Tom Gallagher, a St. Albans cattle dealer, said: "You could do nothing wrong in the 1950s".

And so it was also true during the four years of President Carter.

Quarantined herds were assigned to each of us. We tested each herd once each month and then hot-branded and wrote USDA government permits for positive cows to go for slaughter. If a positive cow was mortgaged, we went to the bank to discover whether the farmer or the bank got the government indemnity check. I tested all fourteen herds in the town of Fletcher with no help. My general rule was that I worked alone as long as I was testing fewer than 100 head.

My first week I went alone to a quarantined farm near St. Rocks in Fairfield. This herd had been under state quarantine for five years. The owner told me no one had tested his cows for over six weeks. He was angry and discouraged. My job was to inform him that the state would depopulate and pay for his whole herd. Then he could start all over again. He was not happy at the prospect of seeing all his cows and heifers "going for beef". So without state or federal approval, instead of moving forward with the government plan of depopulation, I told him I would blood test his cows every two weeks rather than once a month. But I told him he must put dry cows and fresh cows nearest the outgoing end of the barn cleaner and he must vaccinate all his replacement calves. I told him I would dispose of any dead calves in the landfill if he would drag them outside the pasture and barnyard. These were all my own ideas. Only eight months later his herd was free of brucellosis and I was some kind of a hero to this farmer. I never received any acknowledgement from the state or federal government.

After only one month into my federal job, the four of us had to leave for Tiverton, Rhode Island at 2 a.m. due to the discovery of a devastating viral disease called "hog cholera" in a garbage-fed herd of swine. We arrived a little before 8 a. m. About twenty state and federal workers stood around the hotel until 10 a. m. We were finally told to go back to bed! But the next day we worked to euthanize nearly 2,000 pigs of all different sizes out-of-doors all in one day from 7 am to 7 pm. This was the last outbreak of hog cholera in the United States (1975).

Don Brien was a Rhode Island State Inspector. Visiting with him at the task force hotel he spoke about the world's most famous Holstein bull named "Osborndale Ivanhoe". At the Panciera farm Don had to use a step ladder to reach up high enough for drawing a blood sample from the underside of the bull's tail. The best daughter of this bull, named "Woodbine Ivanhoe Mollie" from Pennsylvania, scored Excellent-97 for type. Type means "body build" or shape with the True Type Model being 100. No cow has ever scored over 97. And "the Mollie Cow" produced over 290,000 pounds of milk in her lifetime! That's over 580,000 8-ounce cups! No wonder the dairy cow has been described as the foster mother of mankind. Auctioneers and farmers say interesting things about good cows like Mollie.

"Look at that big udder. This cow is a real mortgage-lifter."

"Her shoulder is so sharp, she could split a raindrop!"

"That cow's muzzle is so wide, she could bite a bale of hay in two!"

"Her back is so wide, you could set a banquet on it!

For a whole afternoon in Montpelier I listened to the state veterinarian and the federal veterinarian argue concerning what to do with a herd of cows with only one brucellosis "reactor". Finally they came to a decision. It was 5:30 pm. I rushed out of the office and whizzed past Beatty's Four Corners to Franklin, getting to the farm at just after 7 pm. I informed the farmer that the quarantine would last four months, so long as no more reactors were found on the brucellosis blood test.

"We'll test your herd once a month. Four months after the last reactor is found, your herd will be released from quarantine."

"I have 40 big heifers sold. Most of them will calve in 2-3 months. What am I going to do?"

"You can milk them, that's all."

"This job is not a popularity contest", I thought to myself.

The next time I was in Montpelier my boss said to me, "Why did you quarantine that herd?"

Dick and I tested 98 cows together at a farm in Swanton well known to him. There was a lot of conversation going on between him and the owner covering many topics. When we went around the barn reading ear tags, we discovered there were 99 cows in the barn!

The farmer asked, "Now what are *you* going to do?"

We simply could not discover our shortfall. I suggested we test the whole herd cows all over again. As the farmer walked away to get his breakfast, we hurriedly recommenced the project. When we got done we had 101 blood samples!

My coworker asked me, "Now what are *we* going to do?"

I answered, "We'll send all these 199 blood samples to the lab with an explanation of what happened. If any are positive we will have to blood test the herd a third time!"

Thankfully the tests all came back negative. After this incident I always instructed the farmer to take some responsibility to keep track of my accuracy and progress. I also purchased a cattle crayon and often marked every 10th cow in the barn once she was tested. Some federal inspectors squirted blood from the syringe onto each cow's backside for inventory control!

Once I went to a farm near Sugar Hill in Derby to blood test a few cows which had been imported from Canada. No one was in the barn. I went to the house and was invited in. I couldn't take my eyes off a briefcase full of cash lying open on a table in the living room.

"What are you doing with all that money?" I asked.

"Oh, I'm going to Quebec today to buy some maple syrup."

I had never seen so much cash, except for the time in about 1960 when a young couple from New York City purchased two family farm properties for $10,000 in cash and third-party checks stacked on our kitchen table.

Several farms in Franklin County sat on the border with Quebec. One farmer admitted to me that he had smuggled cattle into Vermont many times in his life. I'm sure he had learned it from his father! Many people also smuggled liquor and tobacco products, especially during the Depression years. Even today border inspectors have found snowmobiles and farm machinery hidden in loads of hay coming south.

Mom always thought I had big hands and said I would have to work hard for a living. She sometimes called Dad's hands "paws"! In high school the big boys were dunking the little boys' heads in the toilet. They were coming for me, but not before I backed up to the lavatory and held on for dear life. When two boys yanked on my arms the sink came off the wall and water was spraying everywhere. The bathroom was cleared out in a few seconds! I rarely shook a man's hand but what I thought my own hands were as big and strong as theirs. But one day in Fletcher, east of Wintergreen Mountain, I shook hands with a farmer whom I met at his sawmill before going to the barn to test cows. From the backside of the mill came a giant man, perhaps 6 feet 8 inches tall. When I shook his hand it was quite unbelievable. His hands were simply massive, yet soft, each seemingly as big as a couch pillow! Los Angeles Dodgers Manager Tommy Lasorda once described a fellow pitcher's face, rather than his hands.

"His face looked like two rainstorms."

He had hands for pitching, but Tommy thought he'd make a better weatherman than a pitcher!

The Branon and the Howrigan families from Fairfield are folks who know how to work. They each had five or six boys, who together cut and stacked 100 cords of firewood each for the maple sugaring season. They also cut at least five cords of wood for each farmhouse! Tom Branon could buck up a pickup truck load of firewood in no time at all, while visiting at the same time!

One time a farmer in Swanton wanted me to test his cows during morning milking. He said he would be in the barn at 3:30 am. I got up at 2:30 so I wouldn't be late. It was twenty-five below zero that morning! I waited in my car in his driveway from 3:20 until 5:15. As I waited in my warm car many memories of my several "farm jobs" came to mind.

Before I turned age thirteen in 1965 I worked at Whitacres for five summers. My specified hours were 7 am to 7 pm for $25! I worked almost like a man, fixing fence, shoveling gutters every day, raking hay, and more. One day I was in the hay mow stacking hay bales without a break from 11 am until 10 pm.

Fifty-two years later Fran and I were at a barn sale in Barton, when Bryce told me Albert paid him $30 per week after he graduated from high school. He was quite intrigued and satisfied to learn that he made $5 a week more than me! Likely Bryce lived with Albert's family, like I did when I got out of college. Hired men often lived in the farmer's house. There is never much time off. And so it never quite feels like home. But one's clothes are washed and meals are prepared for you. I remember not even having time to go out and buy a candy bar. Of course I would not have dared take the time off, even to go out of the driveway for any reason, for fear of getting fired! Being a herdsman meant you were "tied to the farm", that is, to the land, to the cows, to the weather, to the clock, and to the repetition. This meant three meals a day with no snacks except soda! It is a true adage: "All work and no play makes Jack a dull boy."

On my first college summer I worked for a farmer for only six weeks. He told me to go out at 3:30 to get the cows in for milking. So for six weeks I jumped out of bed every morning at 3:30, threw my clothes on and immediately headed for the pasture, a 30-minute walk away. By the time I got each cow up from the ground and gathered into the barnyard it was 4:45 in the morning. But the farmer kept his clocks an hour ahead of everyone else. So it was actually 3:45! I had already been up for an hour and fifteen minutes. I was working more hours than the owner! Still we must always be willing to work, seeing work as our major activity rather than recreation.

Jesus said: "My Father has been working until now, and I have been working." (John 5:17)

For my senior year at the University of Vermont I lived in an apartment beneath the Animal Nutrition Barn, a research facility. My roommate and I were responsible for taking care of the animals and keeping the barn clean. It would be a month before we received our first check. We survived that month on three things – fresh zucchinis in the garden left behind by the summer interns, the breast meat of pigeons caught in netting we put up near the dairy barn's outside feed manger, and milk taken from the tank after dark!

After college my first farm job allowed me to sleep until 6:30, or so it was originally promised. My job was to finish milking the few cows not milked by the owner and then to do the whole milking in the evening. But the second morning the owner had only milked a few cows and I was stuck with the rest. I finished the morning milking at 10 o'clock. Then on the third morning he got me up at 4:30 to a breakfast of fried heart and buttered toast, and said he had a meeting to go to. I went dutifully to the barn and began my career as a dairy farm herdsman at 5 a. m. Halfway through milking at 7:05 I saw him drive away dressed in a suit coat and tie.

Several farmers used to compete who got up earlier and whose day ended first. One farmer was done for the day at 4:15 pm, but he began at 2:30 in the morning, or should I say "at night"!

Often in northern Vermont farmers have to utilize a bulldozer or crawler to clean their barn. This was because of many repeated nights of sub-zero temperatures. Manure freezes solid in the free stall barns forming huge areas of icy manure throughout the length of the barn. Only a bulldozer can break it up into small enough chunks for the bulldozer to also move them out of the barn.

Uncle Johnnie said Dad used to wear a cardboard box over his head to stay warm when he was bulldozing snow off the roads in Glover! The only two holes in the box were the small ones he cut out for his eyes - another example of Dad's tricks to withstand the cold.

Another thing cold nights would do is freeze the cull carrots from Quebec. Northern Vermont farmers bought them as feed for their cows. They would have them dumped by the barn door and feed a wheel barrow full for every 3-4 cows. The cows loved them. They would swallow them whole or crack them in half. After eating as many carrots as they could, they would stop and shiver all over! Then they would cough them all up, covered in saliva. After a minute or two they would eat the same wet carrots all over again. Only this time they didn't shiver as much or cough them up again. Quebec pea vines are also a popular feed for cows, as is sugar beet pulp, served dry or wet. Cows will also eat squash, pumpkins, and potatoes if they are first chopped up into small pieces with a shovel! And cows will even eat outdated candy!

Milking your cows three times a day became quite popular in the 1980's.

I asked one of the Brooks brothers, "When are you going to start milking your cows three times a day?

"Ha! It's bad enough milking twice a day. If anything, I'll go to milking them once a day."

Having no experience with poultry, I asked a poultry flock owner about how to handle them.

"Do you just pick them up by the hind legs?"

How he laughed! He thought I was joking.

At the Shipman farm my niece told me she had just seen a big pig.

"Did you scratch its neck?"

"It didn't have a neck."

A farmer mentioned to me that his mother-in-law had fought with health problems all winter.

Then he said, "It would be terrible to live all winter only to die in the spring."

On the coldest day I had ever experienced in Pennsylvania I phoned my dad in Vermont early in the morning.

"How cold is it up there? It's three below here.

"I haven't looked yet. Let me look. Oh, it's 47. I've never seen it this cold."

No true northern Vermonter ever says the word "below" when speaking of winter temperatures.

They simply ask, "What did you have?"

The "below" part is understood. When northern Vermonters don't say the word "below", then it's "below" (zero, that is!)

The word "negative" for below zero temperatures is OK to use in the right company, as in "negative 28".

Living near the Canadian border the Davis brothers regularly bought hay out of Quebec. One time the hay dealer said he had a load of very good hay, but had no help to unload it. One of the brothers asked him if he could leave the truck over the weekend and by Monday noon they would have it unloaded. Monday morning the neighbor's goat got loose and was roaming through the farmer's garden. As a temporary fix one of the brothers hitched the goat to the side of the hay truck. After lunch the man came to take his truck back to Quebec. Then he remembered he needed to fuel up before crossing into Quebec. He pulled up to the gas station in Richford and the attendant came out.

"Fill me wi' d' gaz."

The attendant looked strangely at the driver: "Sure, but can you dits-moi me porquois that dead goat is hanging on the side of your camion?"

After a few months of working with the other two state inspectors, I grew wise to their ways. One inspector always arrived at the farm an hour early and bragged about it for two hours. That's because the other inspector was always an hour late! I was the only one that was ever on time! So whenever I scheduled work for the three of us. I gave different times we were to meet to each of them. They couldn't figure out how we all arrived at the same time!

Speaking to a fellow inspector who was prone to continual complaining, Sam chided, "Jim, did you take your 'nasty pill' today?"

My Vermont state cars never had a plug-in heater to warm the engine, and so I religiously added isopropyl dry gas each time I filled the tank with gas from October through May. Yet one time the engine was so frozen it had to be towed to a garage for thawing out over the weekend! Several other times I heated a metal quart can of motor oil on the wood stove in the basement and added half a quart to the engine. It started just like it was summertime!

"You're going to crack the block", a farmer told me.

It never happened. Another trick was to park the car at night headed up to within one inch of the garage door to keep the wind off it. And several times I slid a pan of hot water under the oil pan and opened the hood so the steam could evaporate. I also opened the hood so the sun could shine on the engine, even very early in the morning. Then there was the trick Dad had taught me: bring the battery inside at night and set it on the hot air register. By the way, the best way to keep your feet warm is to put felt (wool) soles inside your shoes. And to stay cool in a car with no air-conditioning, simply put a bag of ice cubes between the seat and your feet. Cy French told me when he started working for the state, cars were ordered special with no radio. A car radio was a luxury unbecoming a Vermont State Animal Health Inspector. But it was one less thing for taxpayers to complain about!

Mike, a cattle dealer in North Enosburg, wisely did not "move" (truck) cattle at all in January. The first month of the year in northern Vermont is plainly too cold for the cattle, the truck, and the driver.

We never butchered on our farm. Dad didn't like it. He said he always ran to the woods when his father and brothers butchered! As a State Livestock Inspector, we dealt with live animals. But about three times each year I had some duties in a slaughter establishment. I always called slaughter establishment co-workers "deadstock inspectors". My first slaughterhouse experience was not as bad as my apprehensions were.

Carl, the deadstock inspector supervisor, had given me a piece of good advice: "When you go in there, put a smile on your face and never take it off!"

My supervisor, the state veterinarian, had given me a list of seven lymph nodes I was to collect as a quarantined cow was being butchered, including mandibular, prefemoral, one-inch pieces of the mammary system called "udder cubes", and internal iliac tissues, among others. Except for "two udder cubes" I didn't even know where these body parts were! But I had a scalpel from testing chickens and away I went to the slaughterhouse. Then I remembered how the Vermont Inspector Supervisor had also told me to tell the veterinarian at the slaughterhouse to collect the lymph nodes for me. I went upstairs to his office. There he was.

After a whole minute he raised his head and said, "What do you want?"

I told him who I was and what I wanted him to do for me.

He said, "I don't need your crap". He stuck his face in his paperwork.

I thought, "What would Dad do?"

So I just stood there for at least two minutes, saying nothing.

Suddenly he jumped up and said, "Follow me."

Down the steel stairs we went and burst through the swinging door, right out onto the kill floor, weaving our way to the starting point.

He barked at the starting man, "Have you seen a brucellosis suspect cow?"

The starting man nodded "no".

We went back into the holding area, found her, and pushed her ahead of all the other cows. Then he told one of the floor inspectors what I wanted and instructed him to "harvest" the tissues on the list. He went back to the office and left me alone!

A worker came over to me and said, "Are you new here?", as he sharpened his knife back and forth on a whetstone practically in my face.

I smiled more and told him what I was doing there. I was always accepted there.

One of the Pennsylvania State Veterinarians, two supervisors above me, called me from Harrisburg with the following information: For years a certain cattle dealer from Maryland had been using the holding pens at a large federal slaughter facility for offloading untested cattle, sorting them and reloading them, all for his own convenience. State and Federal law dictate that all offloaded cattle may not re-enter commerce, but must be slaughtered. This dealer scoffed at the law for years, and would not obey the Veterinary Inspector at the slaughterhouse. One day my boss asked me to direct the dealer to slaughter the 23 young bulls that had been offloaded and reloaded that day. I called the dealer, who was unknown to me. We didn't talk about the weather! I told him who I was and what I wanted done. That's all I said. I didn't tell him why or quote the law to him. I was as brief and as professional and relaxed as I could be. He said he would slaughter the bulls.

"Where?"

He told me the name of the slaughterhouse in New Jersey.

I said "When?"

He said, "Tomorrow."

"What time?"

"By about noon"

I said, "I'll call the inspector right now: Noon tomorrow. Get the ear tags numbers now so I can make out the permit."

"What's your phone number?" he asked.

After hanging up, I called the inspector at the slaughter facility, and told him to call me the next day and send me a copy of the official permit. That was the end of it. I called the state veterinarian back only 10 minutes after he had contacted me. He was astonished!

One time the federal inspectors wrote a permit for a load of quarantined pigs to go to a slaughterhouse. I was instructed to document the movement and disposition of these pigs so that no one took any of them home. I followed the truck from the farm to the slaughter establishment. The truck backed up to the dock. I received the permit, broke the metal seal on the back door of the truck, checked the number for accuracy, and signed the official "Permit for Movement".

About an hour later one of the slaughterhouse owners said to me, "My inspector says all my pigs must be killed and go to the landfill because you broke the seal."

Calmly I told him my job title and handed him my state business card. I told him I had full authority to unseal the truck and had "live custody" of the truck and the animals the whole time. He meandered back inside. Finally the floor inspector came out without introducing himself and let the pigs in. After unloading the pigs I went over to him and gave him the paperwork. If I had stepped inside the building earlier there would have been a kerfuffle, for I wouldn't have been able to say I had live custody of the pigs at all times.

We raised two pigs in our car garage in Swanton two years in a row. The second year someone at the state office let me borrow some brand new butcher knives. I was so tired after butchering the first pig that I had to hire the work done for the second pig! We iced and froze the meat. A farmer told me the meat was all going to spoil because I didn't cure it by "hanging it". But it was the most beautiful meat we ever had. We had picked 700 pounds of corn cobs from the field behind us, but the pigs couldn't eat it because the kernels were so hard and dry! *Agway* had a half-price sale on coarse pig food, so we made out well anyway.

Swanton is an interesting town. The late Queen Elizabeth furnished swans for the town park. Centuries ago England's Royal Navy claimed Swanton for the tremendous white pine trees growing along Lake Champlain half way back into town. By 100 years ago the pines had all been cut down to make masts for England's biggest ships – military and commercial. White pine is light and therefore easy to handle. It would be interesting to discover if any old ship masts of the British Navy ships are marked "Swanton". To this day the majority of carpenters would say white pine makes the best building lumber. Only 6 or 8 trees were left in Swanton by 1900. The forest made up of all these pines is all level farmland today.

One hot summer morning Fran and I were only thinking of getting out of bed. But it was light out. Suddenly we heard a "moo". It sounded like a cow was in our bedroom! We both hoisted ourselves up to a sitting position only to see a cow's face halfway in our widow. We stared back at the cow and said "boo" to her. We looked out the window and saw a dozen "dry cows" lying down under our nearby elm tree, apparently cooling off. We called the neighboring farm and they came and got them after about half an hour. We have no idea how long they had been enjoying themselves away from home.

Several dairy farmers in Franklin County had a barrel of apple cider in the corner of the barn. They added raisins, sugar, ten pounds of beef and a few months of time. The cider turned quite clear and helped chase the cold winter away!

Following are tools and equipment used on the job as an Animal Health Inspector:

Official stainless steel tamper-proof USDA Ear Tags and plier or applicator for identifying cattle and other farm animals and for tracing their movement: Install in the right ear with the point down and the identification number up! Each state has its own prefix number. Vermont ear tags all begin with the number 13. Pennsylvania's prefix is 23. Minnesota is 41 and Arkansas is 71. Hawaii's prefix is 95. Some inspectors mistakenly place the ear tag numbers upside down. Pity the inspector who has to wrestle with a cow's head and ear to read the number the next time!

Knife, scalpel, and spoon for removing the obex (tail part) from the brain of deer for Chronic Wasting Disease testing: A crew of six inspectors sampled "hunter-kill" bucks for two days during hunting season. We trained volunteers, including medical students! The tip of the scapple blade is used to separate the cranium from the spinal column. The blade itself is used to excise (slice through) the dura mater surrounding the obex and spinal cord.

New Syringes and 19-guage one-inch-long needles rinsed, flushed, sharpened, and steamed for obtaining blood for cattle brucellosis testing, and for salmonella testing of poultry: Vacutainers just like the ones used for humans, which are sealed vacuum tubes for obtaining blood via a hub with two needles, one to obtain blood and one to puncture the rubber stopper. California needles are sometimes used. They are Teflon coated, designed to draw blood directly from the neck vein into a glass tube. We also used 5-inch long needles for obtaining blood from a pigs vena cava of the heart!

Noseleads for placing in the nostrils of cattle for head restraint and control, for holding the head in one place where the cow can't hurt you: Northern Vermonters call them "nosegrabs". It gives a cow something to think about while the inspector, the farmer, or the veterinarian are doing other things, such as pulling a cow's infected tooth out or tattooing inside the left ear.

Electric Branding Iron for hot-branding diseased cattle on the left cheek with a "B" for brucellosis or a "T" for tuberculosis: One time a veterinarian plugged in the branding iron and it immediately blew up in his hand. He was badly burned. Red-colored "Reactor Tags" are also applied in the left ear.

Spray rig for sanitizing barns, using a phenol disinfectant: This was a 150-gallon tank with a five-horsepower engine and pump with 100 feet of hose with a nozzle, all mounted inside a cargo van.

We operated another spray rig for spraying cattle infected with scabies, sarcoptic mange mites with a yellow lime sulfur warm water solution. A cow can die from the loss of blood due to intense itching caused by the mites. If you roll the affected skin between the thumb and forefinger, and the cow grinds her teeth and makes saliva, then that's scabies! The Scabies Spray Rig high-pressure hose blew out one time as Charles was spraying the cows. It sounded like a gun going off! It blew the sulfur solution all over his face and then into his boots! Another time Earl was spraying and a cow kicked at him. He had yellow wet lime all over his face!

The state always furnished a car, but rarely a good car. The worst car was a 1978 *American Motors* Concord station wagon. The steering wheel did not line up with the driver's seat! It was off-center by a good three inches! And instead of trading this car after 3-4 years as usual, I ended up driving it for almost 6 years. In Pennsylvania we were provided with the second smallest car made by General Motors. It was small but quite "snazzy".

I hadn't had it too long when a farmer walked up to it and asked me, "Did the state give you a *Volvo* to drive?"

While testing chickens for hours on end, a fellow inspector George said over and over again, "The only good chicken is a dead chicken!"

When there was a slowdown in testing, he would smile but impatiently call out "Birds!"

One time my Mom was impatient with me and my future: "Your father spent his life on the back end of a cow. Does that mean you have to?"

I responded, "Oh, but I spend my time on the back end of a chicken!"

I enjoyed inspecting the St. Albans Commission Sale. John learned from his father "Crick" how to sell cattle at the auction. If a particular cow or bull was heavy, he would introduce her to the buyers by saying "Meat, meat, meat, boys!"

Once when a very thin cow came into the sale ring, a buyer said, "Kinda thin isn't she, John?"

"Thin! She's so thin she'd scare most people! Now who'll give ten cents?"

A farmer once wanted to buy the sale barn.

Crick joked, "You're too honest. You'd never be able to run it."

When the children were young, I developed chronic fatigue syndrome. I used most of my sick leave and took a nap every other day for over a year. We all rested together in our big bed. Our nicknames were "Shove-In", "Shove-On", "Shove-Over", "Shove-It", and "Shove-Down"!

Everyone has had near death experiences or at least dangerous events in their lives.

Two weeks before I was to retire in Pennsylvania, I got deathly sick. Though I made three trips to the emergency room in 24 hours, God delivered me! Our good God is still the main character in all life's stories. My intestinal sickness was from a scratch on the back of my hand from a Muscovy duck we tested at an animal market a few days earlier.

A farm herdsman told me what happened to him when a cow kicked at him as he was walking through the barn. First, he felt a tingling feeling in his jaw, but had no pain. When he reached up to his chin but couldn't find it! The cow had kicked his jawbone off one side of his face.

In 1980 my brother Stuey and my friend Steve rowed a canoe on the Missisquoi River from Highgate Falls to Swanton. We pulled out to get around the five-foot dam in Swanton and continued on our way. About a mile downstream we were forced to decide to go straight or go to the right. We all agreed to go to the right into Dead Creek About a half mile later we were facing the huge bay called "Mississquoi Bay", made up of Rock Bay, Goose Bay, and Gander Bay. Some call this "Frenchman's Bay". As we rowed out into the lake, we made but little progress with a steady 20-30 mile an hour wind in our faces. By sticking my oar straight down into the water I discovered the lake was only about five feet deep. I volunteered to get out of the canoe and walk it across the bay, a distance of at least three quarters of a mile. Though Stuey was a "Navy man", he and Steve strongly protested my plan. But I slithered out of the canoe and stood in water not quite up to my neck!

"Now, when I crawl back in, be sure both of you lean to the other side."

I started walking through the light green lake water north by northwest into the wind pulling the canoe. The whole way the water was exactly the same depth. I was very tired, but finally we were able to turn to the south as the wind drove us to our waiting car. Be sure to study a river map before canoeing!

One year our high school Physics teacher took some members of our high school Science Club to the Boston Science Museum. On the way back I decided lay down on my side in the back seat. Not more than a minute later I felt a terrible thump along with the sound of shattering glass. I was covered in over 100 pieces of glass! My teacher stopped the car and found we had hit another car just on the front corner. We removed as much glass as we could. My shattered window was the only real damage to the car. If I had been sitting up it would have been a whole other story!

Coming home from church in St. Albans a squealing sound started coming out from under the car's hood. When I got home, I opened the hood. I leaned over the engine when I noticed my church tie hanging dangerously close to the fan belt. Just an inch closer and I would have been strangled to death!

Every few years someone was killed by a train which ran between St. Albans and Richford. The train tracks crossed over State Route 105 a dozen times, often with no warning lights. Though I travelled Route 105 almost every day, I was nearly killed on another set of tracks. I was late getting home one afternoon. I longed to get home. I left U. S. Route 7 at Fonda Junction onto a dirt road only two miles from home.

As I approached the tracks I thought: "I'm not bothering to stop".

But in the next instant I said to myself , "You better stop now!"

I stopped within inches of an Amtrak train rushing by the front of my car!

One Saturday while at the University of Vermont I decided to go for a bike ride. I headed off for Shelburne Pond, perhaps five miles away. It was private property, but I didn't care. I walked down a farm lane and through the woods to the edge of the pond. Suddenly a gun went off across the pond. Only few seconds later I heard a bullet fly by my head, turning end over end! I lay on the ground and crawled back to my bike at the edge of the woods. I peddled my bike up the farm lane and made a mad dash north toward the university. I never thought about Shelburne Pond again!

Once I went fishing on the Lamoille River between Cambridge and Jeffersonville. I walked through a pasture and onto boulders that looked big enough to be safe. They had been put there as rip-rap to prevent riverbank erosion. Suddenly the boulder I was on gave way and I jumped onto the next boulder. Five boulders rolled into the river without me!

One time I stopped at the middle of the five-mile-long glacial Willoughby Lake and went down to the water by jumping down on several boulders, each as big as a car. The sky was cloudy and the water was black as tar, being 300 feet deep. When I looked up to the road I realized climbing upon three huge boulders was much harder than jumping down on them! My fear increased when my eyes also beheld what are rightly called "the Willoughby Cliffs" above me at 2,751 feet. I felt dizzy, but I knew I had to defy my fear. I forced myself to breath normally. By reaching up to the top of the boulders and jumping upward I soon found myself back on the road.

Several farmers I knew were killed by their own tractors! As a herdsman on one farm I had to use a gas-starting diesel *Farmall* MD-40 (40 horsepower). It had no push button starter. In fact it had no starter at all! Rather it had to be cranked by hand from the front like an old Model-

T. Then you switched to diesel fuel after starting on gasoline. Think how dangerous this type of machine is! I had never run such a tractor. I wondered if it would blow up. Without comment the owner just assumed I could handle this 1950 monstrosity. I did figure it out all by myself!

Tractors took over for hundreds of thousands of farm workhorses. In the United States alone over 400,000 horses came out of retirement and were shipped on boats to do the heavy lifting in World War I. Ill-fed and abandoned all over Europe, less than 1,000 horses returned home.

Two brothers owned a 200-acre deer farm. They owned wild boars, American buffaloes or bison, and Asian water buffaloes. Several times I was required to walk and inspect the fence. One time the farmer told me he would take me around on his four-wheeler. We were over half done when we spotted an American buffalo. As we got about 20 feet from it, the buffalo shook its head at me and stepped toward us. Patrick goosed the four –wheeler and we escaped.

"Oh, that reminds me. Keep an eye out for the water buffaloes. The engine puttering on this machine makes them want to chase me. This engine sounds just like a grunting buffalo!"

"How fast do they run?"

"Oh, they move right along."

A co-worker got rolled over and over several times against the wall of the "return alley" while standing by a large cow as it was hurrying along after being released by the farmer from the milking parlor.

Our supervisor's comment was, "I don't want any of my men getting hurt."

My boss came out to the farms with me one time in 15 years. He had confidence in all the people he hired.

Once we tested a whole herd of whitetail deer for brucellosis and tuberculosis. I had to go back to the veterinarian's truck for more knock-down drugs and testing supplies. On the way back I heard the rolling of a terrible thumping sound. A big buck with his head down and antlers up was charging straight at me, when he smashed into the chain link fence only a few feet away.

At an elk farm the owner cautioned us never to turn your back on an elk. My co-worker turned to go back to the office as an elk charged at him, crashing into the fence. Several months later a veterinarian was throwing his arms around helping to herd some elk into a pen. One elk came from behind him and only brushed his arm. His arm was broken!

One time a farmer was crushed by a cow in a pasture that tipped over onto him while he was giving her an intravenous bottle of electrolytes to strengthen her after calving. The tired cow got up quite suddenly, but immediately fell over onto him. The only thing that saved him was a passerby who saw his legs flailing in the air and came to his aid. They had to roll the cow off over his head in the same direction she fell on him! He was many weeks recovering.

A farmer always wore a milking stool around his waist. This little chair supported him in a sitting position when finishing the milking process beside each cow. Then one day the cow behind him stuck her leg down his backside firmly into the strap fastened around his waist. The cow thrashed her leg up and down trying to loose herself from the farmer. What a ride! The farmer finally managed to unbuckle the stool. He never wore a milking stool again!

A herd of cows was infected with scabie mites. I informed him how to obtain lime sulfur and then we would spray his cattle at no cost. Instead, he took matters into his own hands. He dusted his cows with alfalfa weevil insecticide. Half his cows died in the barn!

I knew a cattle dealer driver who hid from a mad bull by dashing under the manure spreader! Then the bull got his head under the spreader and tipped it over!

After looking for a missing farmer all day, he was found. His bull had pounded him into the ground such that there was nothing showing but his shiny belt buckle in the mud.

Well, I digress.

During an outbreak of avian influenza in Lancaster County we rode with and advised Pennsylvania State Troopers concerning trucks that might be moving chickens illegally. I worked the night shift eight nights in a row. We were told to never leave the trooper's car. One trooper showed me his pistol at his waist, another pistol strapped to the calf on his leg, and his sawed-off shotgun sandwiched between us.

"You know how to shoot a shotgun, don't you?"

"Yes, trooper."

By the way, never call State Troopers "officers". They are troopers!

One day it was so hot that I went to work with only my bathing suit under my coveralls.

Between farms I stopped on a back road in Fairfield and took my coveralls off; then splashed around under a waterfall. How I'd love to own that waterfall.

Along a high ridge in Montgomery, I pitched down an increasingly steep road in winter, when I realized I had almost no control of my car. I threw the shifting lever into neutral. All I could do was crawl along the shoulder with the right-side tires working the shoulder, hoping not to go into a spin! Sometimes the brakes were helpful and sometimes not. I did a lot of talking to myself!

While passing near Jay Peak before 8 am on my drive all the way to East Charleston, I pulled over onto the side of the road because I heard the sound of a jet. Or so I thought. The roar was actually ice falling off tree branches on hundreds of acres. The falling ice was then rolling down the ice-covered mountainside. The early morning sunrise was just warm enough to melt ice off the trees and cause the whole side of the mountain to roar as the ice rolled hundreds and hundreds of feet. I listened for two minutes. The roar never stopped!

I tested a herd of cows and young stock in the town of Georgia, the next town south of St. Albans. The farmer and I were in the barn together from 9:30 until 1:30 in the afternoon. We were so busy that neither of us noticed what was going on in the outside world. Suddenly we were both shocked to discover 16 inches of snow everywhere we looked; and it was still snowing hard. The farmer offered to plow his long driveway all the way to U. S. Route 7. But my car had four winter tires, and so I went along through the deep snow. Then my car stalled just as I got to the end of the driveway. When I lifted the hood the engine compartment was full of snow to the top! I pulled some snow up and out from around the engine. Then I reached under the engine with my foot and managed to make the snow fall down and out from under the engine. The engine started up once again and off I went.

Franklin County, Vermont had 600 dairy farms when I began working there. Working there and in other counties, I knew 1,000 farmers. Over half of them were French-Canadian. In 2023 there are now fewer than 700 dairy herds in the whole state.

My father attended Craftsbury Academy with the three brothers from Albany. All three brothers became cattle dealers, and their father helped them ascend to the office of bank president.

When I first met Bert, I thought to myself: "What makes him so likeable?'

Then I realized what it was. It was the way he held the phone! He picked up the phone with his wrist bent forward in a very masculine manner. This was an example to me of how to look like a man, at least when on the phone.

A young man at seminary asked me what masculinity meant to me.

I told him, "It's all in the way you hold the phone."

"What are you talking about?"

"Keep your wrist bent forward with your knuckles toward your face."

Being a State Animal Health Inspector caused me to become reasonably good on the phone. We had to schedule all our own work. No farmers called us. We had to call farmers and sell them on the idea of testing their cows.

Horace was one of the most interesting and likeable men I ever met. He delivered mail all around Bakersfield and remembered my wife Franny when she and her siblings were all little ones, "dirt moustaches" and all. This was when they lived in Bakersfield before moving to the farm on Brigham Road in St. Albans in 1966. Horace also bought and sold heifers from Canada and was the most honest cattle dealer I knew. One time I had to appear for jury duty in St. Albans. When I arrived in the courtroom Horace was there along with several farmers I knew. We all had a good time joking around. When the judge appeared in the court room, he asked if anyone had a good reason not to serve as a juror.

Horace jumped up and said, "Your honor, my hearing is not very good."

"Horace, how bad is it?"

"Well, my wife says I miss quite a lot."

Everyone in the courtroom roared with laughter.

Finally the judge said, "Horace, I guess we'll excuse you today."

A farmer from Florida moved 1,800 beef cattle to Shoreham in Addison County. They were all put out on pasture in the fall. We set up a chute and portable metal fence in order to corral them into a place where we could blood test them. Thankfully my co-worker did all the collecting of blood samples. It was a marathon session every day for nine days! For "energy feed" some huge Chianina cattle were fed outdated Mounds and Cadbury candy in the roughage mix. Their heads and ear tags were all sticky! Chianina are one of the oldest and some of the biggest cattle in the world. The heads of the females are as big as the males of most breeds. They originated in central Italy centuries ago and are a challenge to work with!

Buffalo are also a lot of work to handle and work safely around. The first buffalo herd I worked with had a cow which tossed the heavy iron stop gate up out of the chute, by hooking it with one horn and suddenly throwing her head up. She sent it flying ten feet up into the air before it landed beside me! One time I tested a yak. It was staying in its own air-conditioned hut on the edge of a pasture! Yaks cannot tolerate temperatures above 80 degrees.

Another co-worker, used to own a dairy farm. Because he heard they were more efficient and profitable, he changed over from large Holsteins to smaller Jerseys.

But he complained, "I used to have 50 problems. Now I have 70 problems!"

A certain farmer called me "Mr. Brucellosis" because of my years of experience. One farmer called me a "vampire"! I tested over 50,000 head of cattle and 100,000 head of poultry in Vermont from 1975 to 1992. I also sifted and sorted through 20,000 milk samples twice each year to find 2,000 Pennsylvania farm samples for the "ring test", a quick screening test for brucellosis. An antigen for brucellosis is added to raw milk. Even if only one cow in a herd of several hundred head of cattle produces antibodies, those antibodies "clump" together. Over a period of hours these clumps rise into the "cream layer" forming a blue ring at the top of the tube of milk. Now you have a "suspicious herd".

When we discovered a true brucellosis reactor, then we also placed the farm under a state quarantine and posted a large yellow sign for a minimum of 5 months. We tested and slaughtered many cows and were able to clean up some of about 60 infected herds. We began testing all of Franklin County in 1975 and were completely clean by 1980. The heavy-duty plastic yellow quarantine signs made a great cover for my car radiator when it was negative 38!

We did depopulate a few dozen herds. Once the federal and state veterinarians and the farmer knew there was no hope of overcoming the disease, then the decision was easy. They all had to be slaughtered. The cattle were appraised for value, hot-branded with a "B" on the left jaw, and permitted for slaughter. Thirty days after we disinfected the empty barn, the farmer could repopulate.

Several times a year I helped Dr. Krause test cows for tuberculosis. He was the veterinarian stationed at three American ports of entry with Quebec, including Derby and Highgate in Vermont and Rouses Point in New York. After he had injected the first 5 cows, I would begin identifying the cows by reading their ear tag numbers or putting in new ones.

Often he said to me, "No one can get TB testing done like us".

Soon he would say: "Let's go get a sticky-bun, Phil."

My boss asked me to attend the State Ayrshire Sale at the University of Vermont farm arena in Burlington and write interstate health charts. I told him I first had to write a permit for a shipment of positive brucellosis cows in Derby to a slaughter establishment in Pennsylvania.

He said, "Well, you can do that, but be in Burlington by 1:00 o'clock."

So I left early and rushed up to Derby. Once the job was done I hurried to Burlington. It was 20 minutes after 1:00 o'clock when I walked into the arena.

Suddenly the auctioneer stopped the sale and yelled: "Are you the man from the state? What's the answer to that question?"

Over two hundred people, mostly farmers, turned and stared at me.

I hollered back, "What question?"

Most of the audience laughed profusely while the auctioneer went into a public tirade against me and the state. I went out to use the UVM farm office telephone and left a message with my boss; then went back into the arena to write the health charts for interstate movement. Ten minutes later my boss called and said that no one but Vermont farmers could purchase cattle from Orleans County farmers! I went back into the arena. I motioned to the auctioneer, and he stopped the sale again.

"Folks, we have an answer to our question."

With them all turned around and staring at me again, I said loudly, "No Vermont cattle from Orleans County can be sold to out-of-state buyers today."

The auctioneer bellowed: "Isn't this just like the state, to tell us nothing until we are halfway through the sale!"

Several farmers came over to me and started barking into my ear and flailing their arms at me.

The auctioneer never once introduced himself to me or even knew my name. It was not much fun for a 25-year-old.

I arranged to obtain a milk sample for brucellosis testing from the bulk tank at a farm in Starucca, Pennsylvania.

The owner said, "I won't be there, but help yourself."

Every bulk tank is different. When I entered the milk room I examined the controls to be sure I understood how to agitate the milk before obtaining a milk sample. It was a pleasant winter day and I was enjoying waiting for the milk to agitate on a timer. After the agitator shut off I opened the front cover. I dipped into the tank and poured the milk into a plastic vial. Plop! I dropped the vial into the tank! The tank contained about 10 inches of milk. So I washed my hands and wrists and reached down over the tank and into the moving milk. The vial was not there! I went to the other end of the tank and opened the back cover. Reaching down into the milk I found the vial lying in the lowest part of the tank. So, there I was all covered in milk when I heard the sound of someone outside walking on the crunchy cold snow toward the milk room!

I thought, "He's going to ask why both covers are up on the tank."

I quickly shut the back cover and went over to the sink and was rinsing milk off my arms when the milk room door opened. I pulled my sleeves down, put the vial in my pocket, and pulled down a paper towel.

"Good morning, Bob."

I turned around, shut the front cover, and wiped off a few droplets of milk. After a good visit, I left and found a good cup of coffee.

The Fence Farm near Belvidere Junction in northern Vermont had a big Jersey bull in a pen. The farmer hitched a rope to the bull's leather bridle, and I entered the pen. When I grabbed the tail to draw a blood sample, the bull switched his rump to and fro and I could barely lift the tail. It had been broken long ago and had healed in the down position.

I came out of the pen and said, "Let's give him something else to think about."

We put a rope halter on his head and hitched it tight in the opposite direction to the bridle. Then I put "nosegrabs" in his nostrils and asked the farmer to hold on tight. Back in the pen I took hold of the tail. How he bellowed! I got the tail up just enough to get the blood sample from the underside of the tail.

"Huh. That's the first time my bull ever got tested. The veterinarians always just draw blood from a cow instead."

Sometimes a strong farmer can grab a cow's nostrils with his bare hands and hold their head still. A strong man can also stand beside a cow and squeeze her heartily at the chine, the front of the back along the spine, making the cow "shrink" and act quite submissive!

A farmer once told me of a bull that got him down in the free stall barn, placed his horns around his ribs, and pushed him up and down the wet barn floor, bellowing over and over again. After five trips back and forth on the wet floor, a cow bellowed. The bull picked up his head. The farmer lurched up and started climbing the rafters with the bull shaking his ahead and bellowing harder. With the bull following him all the way, he climbed along the rafters to the area of the door to the milk room. After a long time, the bull meandered off and the farmer jumped down and dashed into the milk room.

A farmer near Cleveland Hill in Coventry refused to separate the cows I needed to test. Instead, he chased bunches of cows from the herd into corners of the barn and haphazardly identified certain cows he had recently purchased from Canada. Finally, the majority of the herd stampeded from the new section of the barn into the old barn. In the process of "moving as one", the cows on the outside of the herd knocked a whole wall out of the barn. A dozen huge hand-hewn cedar rafters bounced up and down. Thankfully none of them collapsed or broke off!

Another farmer told me how his cows stampeded out of the barn when a veterinarian yelled in the barn, "Is anybody home?"

Again, it made for another repair of a barn wall. Farmers always have plenty of work – regular and repairs!

Five of us were testing cows at a large farm near Sanderson Corner in Fairfax. After getting the milking cows tested, we were standing around visiting while the famer was getting the dry cows rounded up for us. Dr. Powers, who was ready to retire, disappeared. He had walked out into the barnyard and began petting a young bull when we heard him hollering for help. A young bull had knocked him onto the ground and was head-butting him.

After we all rescued him, someone asked: "What were you doing in there?"

"I just wanted to pet that young bull."

Of course, prejudicial sayings about the Northeast Kingdom are not well-known outside northern Vermont itself.

I've heard it said many times: "There's not much north of St. Johnsbury."

In fact, the Northeast Kingdom boasts the most fabulous displays of geography in the state. Willoughby Lake with its rock cliffs is a National Natural Landmark (*NNL*). Lake Memphremagog in Newport extends over 25 miles into Quebec. Every spring salmon can be seen jumping up the falls of the Willoughby River in Orleans village.

While at UVM, a teacher said to me: "It's a wonder you can even talk".

At the time I knew nothing of such prejudice, and so I said nothing in return, thus reinforcing the stereotype!

Forty years later I heard someone repeat these exact same type judgmental words on a radio talk show. NEK author Howard Mosher experienced the prejudice other Vermonters have against others they assume to be "backward", or at least a generation or two behind the times. This prejudice reminds me of a 400-year-old dictionary which gives the definition of the word "oats" as a grain the Scot's eat, but the

English feed their horses! My modified tale told by Francis Colburn pictures this attitude:

Two men were making counterfeit $20 bills in Rhode Island.

When they were done they said, "Let's make sure this works. Let's start by going to northern Vermont first before using these $20 down here. Vermonters up there are pretty backwards."

"Well then, let's make an $18 bill and have them cash it."

So off they went to Greensboro in the Northeast Kingdom. In the little store one of the men bought some cigarettes with the fake $20 bill with no problems. Now feeling more confident the other man asked the storekeeper to exchange his $18 bill for some smaller bills.

After a little pause the backward storekeeper replied, "Would you like three 6's or two 9's?"

You can adapt this story to any state!

Philip Alexander Urie

7

Farm Dogs

The final chapter of stories is about country dogs, not city dogs, specifically *farm dogs*. Although we expect our pet dogs to live to age 14 or more, farm dogs live to be only 7-8 years old on average. They work too hard and are exposed to too much cold weather.

My mother once said to a friend who stopped to visit with her after many years of absence, "I haven't seen you in a dog's age." So it was likely 10 years since her previous visit at least.

Farm dogs are more "guard dogs" rather than pets or friends. A farm dog rarely has house privileges unless his name is "Mutley". But a dog is something that a farmer needs outside and puts up with in the barn. A farm dog will bark or growl at the sign of anything unusual. Farm dogs are appreciated almost like a real person, especially if the dog can herd animals to where the farmer wants them to go faster than the hired man. Most farmers talk to their dogs when they think they are alone with them.

A good day is worth a lot. Steven Spielberg's animated film about "Balto" relates the true story of the heroic delivery of life-saving medicine through an Alaskan blizzard. Just before he died in a plane crash in Alaska, writer Will Rogers penned these words: "The backbone of the Artic is the backbone of the dog." Jesus is the backbone of our strength.

He said: "Therefore my Father loves me, because I lay down my life that I may take it up again. No one takes it from me, but I lay it down of myself. I have power to take it again." (John 10: 17-18)

Farm Dogs

From 1961 until 1966 a dog named "Prince" ruled West Glover village. We had to face him every time we walked to the store to buy penny candy. He never once growled or bit anyone, but he possessed an ominous charisma. He filled the village by his very presence. He was big and pranced everywhere he went. One day I stayed home while everyone else went to Newport to go shopping. Two minutes later I saw Prince prancing down the road looking especially bold. Another dog I had never seen accompanied him. Suddenly I remembered our farm dog was sleeping on the front lawn! Prince walked up to our dog and shook him furiously; thankfully Prince shook him only one time. Our dog was never the same. No one ever dared to confront Prince or his owner.

As a State Inspector in Franklin County, I learned a lot about befriending dogs with food. I also learned that a dog could be disarmed by talking to them in a child's high voice and with children's words, as in the word "doggie" rather than simply "dog"

On my first week of work a bulk tank driver said to me about a very protective farm dog: "Feed that dog your sandwich after you take a bite in front of him."

So, I always carried a bite or two of leftover sandwich and a box of dog biscuits in my car.

Once, while disinfecting a barn, I left the barnyard gate open. "Phil, the cows are in the corn1" I thought to myself, "Are the sheep in the meadow?"

Thankfully the farmer's dog herded the wandering cows back inside the gate.

A farm I had visited many times suddenly had a particularly aggressive new farm dog which would surely have bitten me if I had turned my back on him. He would not let me into the barn, even when I swung my boot brush at him.

After the dog almost attacked me at the back of my car, the farmer peaked out of the barn door and said, "No one's going to steal tools out of my barn again!"

Then he laughed and shut the door after the dog went back inside with him. So, I went into the milk room and disinfected my boots. When I walked into the barn the farmer had to bark at the dog quite several times to keep him away from me. The dog barked and growled at me the whole time I was testing three cows. The next time I went to this farm I gave a ham bone to the dog. The dog was quiet as a mouse. When I squatted down to empty my syringe of blood into a tube in my kit on the floor, the dog licked my face! He never barked at me again, maybe because I always brought a ham bone for him!

When I pulled up to the barn at the Rene Cyr farm in St. Albans, I noticed a dog on a chain near the barn door.

Suddenly the owner came rushing out of the milk room, saying, "Back up, back up, don't park here!"

I rolled down my window and asked, "Why?"

"My dog bites holes in inspector's tires!"

He elaborated once I got in the barn: "My dog flattened both front tires on the milk inspector's car."

A veterinarian and I had to collect nasal swabs at a horse farm. As soon as we arrived, I got my paper coveralls and plastic boots on.

Then the barn door opened, and I heard these words: "Is anyone afraid of dogs?"

At almost that same instant a big Shiloh (white) Shepherd came running out of the barn straight for me, barking and growling with pure animal abandon. The dog stuck her nose straight up inside my rump and growled ferociously. We went inside and swabbed several horses while this dog barked and growled at us the whole time! When I got back to the office I told a State Dog Warden about this "dog event".

He said "Oh, I know that dog. That dog bit several people. That dog is officially a "dangerous dog". It's supposed to be on a leash at all times."

I had never been to a certain farm in the Hectorville neighborhood of Belvidere. When I pulled in I noticed a German Shepherd lying on the porch. I waited five minutes when finally I noticed the dog was no longer there. So I got out of my car and headed for the back of the car to get my supplies.

There he was, growling at me, hair on end, standing at the back of the car, when I heard these words from the milk room window: "Don't you turn your back on that dog."

"Don't worry." I said. "I won't."

"That dog's put three men in the hospital. One man came here drunk and said he wasn't afraid of my dog. Well, let me tell you, he sure found out who the boss is. He jumped back in his car right through the window head first!"

The whole time I was there I followed the owner everywhere he went, and the dog followed me!

"One of the vets kicked my dog in the chops and he doesn't like men in blue coveralls."

This dog must have been kicked for a reason! He was mean long before any veterinarian had kicked him. Two years later this farm dog was run over by the neighbor's tractor.

But the worst dog of all was the one I never met! I was out investigating a dead horse complaint. I had seen a "Beware of Dog" sign at the farm and something inside caused me to decide not to enter the premise. I drove by the place three times with every intention of entering but did not. Instead I stopped at the post office to enquire about the farm. I asked the postmistress if there was anything unusual about the farm.

"Are you kidding me?" she said. "It took the State Police three weeks to dig that man out of a cave!"

I decided to do nothing! Instead I contacted the township supervisor, who said to me: "I'm glad you didn't go there. My wife is the Town Health Officer, and that dog attacked her bad. She's been in the hospital for weeks."

When I got back to the office I told the Dog Warden people about it.

"Oh yes, we know about that dog. It's been officially declared a dangerous dog twice, but none of us have gone to pick it up yet."

One cold fall day I went to Alburg to test some cows. I couldn't find any contact information for the farm and so I asked about the farm at the post office.

The counter person said to me: "You aren't going there, are you? He's got one bad dog and the other one is almost as bad. He rents his barn to the man you are looking for."

I drove up the long driveway as far as I could. A truck was parked blocking the way to the milk room. I waited 20 minutes just sitting in the car.

Finally I said to myself, "Well, it looks safe enough."

I took my supplies in my left hand and carried my boot brush in my right hand for protection. I walked halfway to the barn when two German Shepherds came bolting out of the house! I smoothly moved the brush alongside my thigh to hide the fact that I had a "weapon" in my hand. One dog sat down beside me and did nothing but glare at me and growl. When I spoke to the dog in a friendly way using a high child's voice, he could see right through me. He just growled and dared me to move an inch. The other dog kept moving all around me in a sweeping motion. After about two minutes the dogs ran back toward the farmhouse, and I watched them being let back inside the house. When I got into the milk room my knees buckled under me with no strength.

After testing four cows I said to the farmer: "Please go with me back to my car."

He said, "Oh, he won't turn them out on you again."

A large animal veterinarian raised German Shepherd dogs on the road next to us in Swanton about a mile away. One time I told him how I didn't like a certain German Shepherd at a farmer's "home farm" in East Enosburg.

He said, "Phil, don't let that old farm dog bother you. I went there last month, and nobody was in the barn. As I walked to the house he riled up at me. So I gave him a heavy shot of pepper gas. He went down and took care of his face, let me tell you! No one answered at the house. So on the way back to my car, I blasted him with another shot."

One winter's day at a farm near Willoughby Lake I had to fight my way into the milkhouse. My fight was with a goose. It flapped its wings ferociously at me and lowered its head in an attempt to intimidate me. It wanted me to turn and run so it could pinch me on my backside with its bills. After I tested three cows and had washed my boots, I asked the farmer to go out and scare the goose off.

"You're a fraidey-cat."

The farmer opened the door, and there he was - the goose in all his masculine finery and pomp - right in front of the door with his wings spread wide! At the sound of his honking the farmer slammed the door in the goose's face. He found a stick and swallowed his fear. Then he opened the door again and burst outside sounding louder and more aggressive than the goose. Off it went into the field honking and honking.

"Thanks. I didn't want to do that to your goose."

Several different times I walked to the backside of Metcalf Pond in Fletcher for my lunch break. It was "wilderness country". Once I heard a scratching sound behind me. It scared me to no end! Without looking back I ran full tilt for several minutes to get back to my car. Perhaps it was a stray dog, but more likely it was a bear smelling my tracks!

In 1972 Stuey and Mikey adopted a stray dog and named him "Mutley". After eating his meal of leftovers Mutley tried over and over again to use one of his paws to scratch out a chicken bone which was caught in the roof of his mouth. When the boys saw him working hard, they informed Dad. He quickly brought a pair of pliers to grab onto it and remove it. But when Mutley saw Dad coming, he used *both* paws and hurriedly removed the chicken bone all by himself! Perhaps he remembered having 20 porcupine quills taken out of his nose by Dad and me a few years earlier. The dog we thought had become our farm dog had abandoned farm life, choosing instead to run in the fields and woods, following deer tracks for hours at a time. He decided he'd rather spend his time hunting than hanging around the barn. Mutley had become a dedicated hunter! So Dad dangled a lightweight chain from his collar and attached a 6-inch piece of round wood cut off from an old house broom handle. The small piece of wood knocked lightly on the knees of his front legs. Three days later he was full of quills and too tired and sore to hunt deer.

Another time Dad removed Mutley from the house with only a look! The boys had taken their farm dog into the house to watch *Batman* together on the television. But he left the house with his tail between his legs once Dad came into the room. Dad didn't say a word to his farm dog!

Dogs are said to be "man's best friend". But this does not mean that every dog is everyone's friend. When in 1933 the man from the *Federal Land Bank* came the final time to tell the Alexander family they had only a week to get off the Aldrich farm, Wayne sicked their farm dog on him. The government man got bit in the buttocks! Not every day on the farm is magic!

Author Bio

Philip Urie was raised in the Northeast Kingdom of Vermont in West Glover village. His wife Frances was raised in Bakersfield and St. Albans. They raised their three children in Swanton. Phil graduated from the University of Vermont in 1974 and earned a master's degree from Reformed Presbyterian Theological Seminary in Pittsburgh. He served as a church ruling elder.

Roderick Wells (1931-2019) was a professional illustrator, artist, and teacher, who earned degrees at Columbia University, where his father taught English literature. His illustrations appeared in books and nine magazines, including *House Beautiful, Gourmet Magazine*, and *The New Yorker*. His work has been aptly described as "Romantic Realism". Rod was the 7th generation of the Wells family to own land in Danville, Vermont.

*A*dvantage
BOOKS

advbookstore.com

www.ingramcontent.com/pod-product-compliance
Lightning Source LLC
LaVergne TN
LVHW021612080426
835510LV00019B/2534